Horticultural Export Management

Horticultural Export Management

Jayati Vijaywargiya
Meera K
Shailesh Prasad

2026
Agri Horti Press
New Delhi-110 084

ISBN: 978-93-49690-59-2 (Hardbound)

Published By

Agri Horti Press

New Delhi

E-mail: agrihortipress@gmail.com

Digitally Printed at : Replika Press Pvt. Ltd.

PREFACE

In the era of global trade and rising consumer awareness, horticultural products have emerged as high-value commodities with significant potential for international markets. Efficient and strategic export management of fruits, vegetables, flowers, and other horticultural goods is essential for maximizing market access, ensuring quality compliance, and enhancing the global competitiveness of producers and exporters alike.

This book, titled *Horticultural Export Management*, is designed to provide a comprehensive overview of the principles, practices, and challenges involved in the export of horticultural products. It serves as a practical guide for students, agribusiness professionals, exporters, policy-makers, and others engaged in or aspiring to enter the global horticulture trade.

The book begins by outlining the fundamentals of export management, including export procedures, documentation, and legal frameworks. It explores the dynamics of international trade, market selection, and entry strategies tailored to the unique needs of horticultural products, which often demand strict quality, safety, and shelf-life standards.

Horticultural Export Management covers critical aspects such as post-harvest handling, cold chain logistics, packaging technologies, and compliance with international phytosanitary and food safety regulations. Special emphasis is placed on understanding global trade agreements, certifications, and the role of organizations like APEDA, WTO, and Codex Alimentarius in shaping horticultural export policies.

Real-world case studies and industry examples are integrated to provide practical insights into overcoming trade barriers, managing risks, and leveraging export incentives. The book also addresses sustainability issues, emerging trends, and digital tools that are reshaping the horticultural export landscape.

By combining theoretical foundations with actionable knowledge, this book aims to equip readers with the skills and understanding required to navigate the complexities of horticultural exports. Whether the goal is to expand market share, improve export readiness, or contribute to agricultural value addition, *Horticultural Export Management* offers a valuable foundation for informed decision-making and global success.

We hope this volume will serve as a reliable resource for those committed to advancing horticultural trade and promoting excellence in export management.

Jayati Vijaywargiya

Meera K

Shailesh Prasad

CONTENTS

1

INTRODUCTION

Export is a vehicle of growth and development. They help not only in procuring the latest machinery, equipment and technology but also the goods and services, which are not available indigenously. Therefore, it occupies a very prominent place in the list of priorities of the economic set up of developing countries because they contribute largely to foreign exchange pool. Export plays a crucial role in the economy of the country. In order to maintain healthy balance of trade and foreign exchange reserve, it is necessary to have a sustained and high rate of growth of export.

A product that is sold to the global market is an export, and a product that is bought from the global market is an import. Import allows us to access global products and services at competitive price. It makes possible to access global technology for up-gradation of indigenous production technology, better education, better health and better transport services that otherwise may not have been available to us. This ultimately leads to a better life.

Whereas, export allows us to expand our markets at global level. It allows countries to use their resources - whether labour, technology or capital - more efficiently. Because countries are endowed with different assets and natural resources (land, labor, capital and technology), some countries may produce the same good more efficiently and therefore sell it more cheaply than other countries. If a country cannot efficiently produce an item, it can obtain the item by trading with another country that can. This is known as specialization in international trade.

International trade not only results in increased efficiency but also allows countries to participate in a global economy which raise employment levels and increases the income of individuals and nation as a whole. This leads to stimulating economic growth.

Global Horticulture Market Outlook 2015 projects that horticulture industry plays vital role in the future. The globe has initiated measures to support this industry. In 2011, global fruits and vegetables production were 548 million tonnes and 990 million tonnes respectively. And the global floriculture industry size was around USD 109 billion.

Globalization has witnessed the focus on integration of developing country firms geographically with PAN world with supply networks or commodity chains. These global supply chain linkages help to connect developing countries' firms with developed countries' suppliers and customers. Gereffi, emphasises not only that independent companies in different countries are linked together in trading relationships, but also that the chain should be considered as a network governed to a large extent by key agents within it. Watts's paper focused on the trade in fresh vegetables and said that this sector has become one of the most vibrant sectors in international trade, and during the l990s, imports of fruit and vegetable products by EU countries surpassed all other categories of agricultural products (Wattsl 994).

Gereffi reported GCCs [Global Commodity Chains] novel in increasingly organizing international production and trade in strategic decision making and economic networks at global scale by industrial and commercial firms. Gereffi classified the commodity chains are producer-driven and buyer-driven. Producer-driven commodity chains are governing by transnational firms and buyer-driven commodities govern by retailers, importers and brand-name companies. These firms link overseas factories and traders with evolving product niches in their main consumer markets.

India is one of the largest agrarian economies of the world. In the context of economic liberalization, it promotes export led growth. The Indian National Agriculture Policy (NAP) focuses more on demand driven agricultural growth in domestic and export markets. Of late, WTO trade liberalization and multilateral negotiations also focus more on international agricultural commodity markets in the light of improvement of market access and limitation of export subsidies.

Horticulture marketing define as finding out what your customers want and supplying it to them at a profit therefore involves: identifying buyers, understanding what they want in terms of products and how they want to be supplied, operating a production-marketing chain that delivers the right products at the right time and making enough profit to continue to operate.

Horticultural sector contributes 54% of agricultural Exports and 28% of AgGDP in India. Over the last decade this sector grown at the growth rate of 3.6 percent. NHB, 2011 reported that India is second largest producer of fruits and vegetables in the world next only to China. This tremendous production potential offers India's greater opportunities for exports. India produced in 2010-11, 146.55 MTs of vegetables and 74.88 MTs of fruits. And its requirement is 175.2 MTs vegetables and 74.40 MTs fruits. In 2011-12, India exported vegetables and fruits worth Rs. 4801.29 crores. The consumption basket is changing towards fruits and vegetables. National Horticultural Mission started in 2005, Foreign Trade Policy (2004-09) advocating to boost export growth and promotion of horticultural products. XI & XII Plans also emphasizing accelerated horticultural growth.

India is one of the world's leading producers of finest tea 23 per cent share by volume in 2013. Its contribute 12 per cent share in world tea exports in 2013-2014. The CIS countries accounted for 19.8 per cent of total exports in 2013-2014 with US$ 152.46 million. India registered exports of 225.7 million kg. for approximately 78 per cent of total exports during 2013-2014. In 2013-2014, the Russian federation was the topmost importer of Indian tea, followed by UAE and Iran respectively.

Coffee Board, 2014 reported that India ranked in coffee production seventh largest in the world and third largest Asia. India exported coffee during 2013-2014, 313,025 metric tonnes and earned foreign revenue 793.22 US$ million. India exports coffee to over 45 countries, over 50 per cent of to Europe. Italy is the largest importing more than 25 per cent from India.

We are living in multi-speed world economies. Horticultural commodities play significant role in living economics. World Development Report reported that markets are central to the lives of poor people and World Bank, World Development Report framed institutions that support the development of markets.

UNCTAD study revealed that India's exports to world are income elastic and found that 1% decline in GDP growth of world will lead to 1.88% decline in India's growth of exports. Singh and S. K. Singla found from their study (1992 to 2006) that India has been unable to diversify its exports as well as export destinations. Export from India has experienced instability primarily in the developing countries. Demand for India's exports was found to be elastic to both relative prices and income of the importing countries. Traill predicts that by 2015 supermarket's share of the retail food market will reach 61% in Argentina, Mexico, and Poland, 67% in Hungary and 76% in Brazil. IFPRI's research shown that agricultural Market reforms liberalization programs adopted by many developing countries in the past had limited success in developing private, efficient, and competitive agricultural markets. Dastagiri and Immanuelraj advocates reforms needed to shift Indian farming from traditional farming to an agribusiness sector and develop linkages farmers to super markets and export markets is a key driver for industrialization of agriculture.

Alberto and Stefano estimated import price demand elasticity in the destination markets of Italian exports found that Italy has a relatively low elasticity of substitution in main specialization sector (machinery and equipment), while has higher substitution elasticity for traditional goods like textiles, jewelry and leather. Reardon and Hopkins ; Minton (2008) found that super markets and retailers competing om price, convenience, quality of the products and safety. Mithofer et al. found that smallholders were linked to the export market by middlemen in Kenya. Galanopoulos et al. reported that though Mediterranean countries are traditional growers of fruit and vegetables but unable to competitive in the global market. Dastagiri et al. estimated and found that for all vegetables the NPC is less than 1 indicating they are competitive in the international markets. Lenné & Ward found that the export vegetable subsector as a role model for improving the efficiency, growth and economic value of domestic vegetable marketing systems in East Africa. Dastagiri et al. reported that most of the Indian horticultural commodity markets were operating efficiently and producer-consumer was the most efficient channel. Diego et al found in Niamey, Niger for market retailers net profit (NP) for amaranth and lettuce on marketplace, whereas the NP of cabbage and tomato strongly depended on season and marketplace.

This paper examines the position of Indian producers and exporters in the horticultural export trade from the perspective of the global supply chain analysis. There are limited studies on market research of horticulture commodity global supply chains and identification of major global markets. This study analyses Indian horticultural commodities export markets linking to global supply chains and price trends, elasticity's and instability & identification of their destinations. Finally, the study will suggest multispeed strategies for improvement of global supply chains, prices, production, and trade.

The Specific Objectives are:

(1) To analyze exports price growth rates of India's horticultural commodities.

(2) To estimate export elasticity's for Indian fruits, vegetables, flowers and Tea & Coffee.

(3) To identify major destinations for horticultural commodities and their share in world horticultural exports.

(4) To trace India's horticultural commodities Global Supply chains.

(5) To suggests polices and strategies for boosting Indian horticultural exports and foreign earnings.

BACKGROUND

At the outset, it needs to be clarified that horticulture as a discipline of agricultural science encompasses study of large groups of fruits, vegetables, mashrooms, flowers plantation crops including cashew, spices, medicinal and aromatic plants. India is rich in diversity of these crops due to immense agro-climatic variations, enormous bio- diversity, fertile soil, large cultivable area and above all long history of crop husbandry. The rich cultural diversity of this country has further contributed to the planned exploitation of crops and trees, giving rise to a large variety of culinary recipes. The total area under these crops in 1994-95 was 14.5 million hectares with an annual production of 119.2 million tonnes. Fruit and vegetable together contributed 90.2 per cent of this production and 65.8 per cent of the total area. (Kaul, 1997).

Table 1.1 Area and Production of Horticultural Crops in India.

(Area in lakh ha. Production in lakh tonnes)

Commodity	1984-85		1994-95	
	Area	Production	Area	Production
Fruits	25.40	237.60	35.71	388.35
Vegetables	58.00	608.80	59.70	686.82
Spices	16.78	12.70	24.01	24.66
Coconut	11.90	44.57	16.90	85.62
Cashew	5.02	2.11	6.35	4.18
Arecanut	1.87	1.92	2.35	2.72
Total	118.97	907.70	145.02	1192.35

Source: Production Year Book, 1997, NHB Gurgaon.

India has a vast potentials of export oriented horticultural crops but due to variation in agro-climatic conditions different fruits and vegetable are grown in different states. Therefore, study is concentrate only on certain crops and upon certain areas which has potentials for growth of concerned export oriented crops. So, the different crops in different states have been identified and studied separately.

Table: 1.2. Selection of Export Oriented Horticultural Crops in Different States.

Name of the State	Export Oriented Horticultural Crops Identified
Andhara Pradesh	Mango,Grapes and Onion
Bihar	Litchi
Gujarat	Chiku(sapota)
Haryana	Cut-Howers
Himachal Pradesh	Apple
Kerla	Chashew Kernels
Maharastra	Graps and Onion
Sikkim	Orchid
Tripura	Pine Apple
U.P.	Mango and Potato
West Bengal	Orchid

The detailed state wise scenario are as follows:

ANDHRA PRADESH

Andhra Pradesh ranks second in area and production of fruits and other horticultural crops in the country next only to Uttar Pradesh with 11.43 lakh hectares of which 5.98 lakh hectares constituting about 52.10% under fruits, 14.72% under vegetables including onions, 2.00 under flowers and 31.18% under spices. Among fruits, mango occupied a predominant position with 23.00% of the total area under horticultural crops in the state, followed by cashew with 10.45%, coconut with 7.84% and citrus with 5.50%. In the case of production mango also accounted for the highest proportion of 35.97% of the total production of various horticultural crops followed by banana with 11.20% and citrus with 10.72%. among the vegetables, tomato occupied predominant position with 4.18% followed by onion with 2.09% and brinjals with 1.66% of the total area. In the case of production, onion shared the highest proportion of 6.83% followed by tomato with 5.42% and brinjal with 4.30% of the total production. Among the spices, chillies, turmeric and coriander are the important crops accounting for considerable proportion of the total area under horticultural crops during 1995-96 in the state.

A hundred per cent increase in the area under Mango from 1.3 lakh hectares during 1970-71 to 2.6 lakh hectares during 1995-96 has been recorded in the state. Similarly, production has also increased from 16.07 lakh m.tons during 1970-71 to 31.64 lakh m.tons during 1995-96. In the case of grapes also, a similar increase in the area from 1309 hectares to 2315 hectares and production from 7120 m.tons to 50,895 m.tons has been recorded during the same period in the state. Area under onion has also increased substantially from 13,396 hectares during 1970-71 to 24,034 hectares during 1995-96. However, a multifold increase in the production of onions from 1.10 lakh m.tons to 6.01 lakh m.tons is recorded during the same period in the state.

BIHAR

Bihar is one of the important litchi producing states of India. In Bihar, it is mainly cultivated in the districts of Muzaffarpur, Vaishali, Sitamarhi, East and West Champaran, Samastipur and Darbhang of North Bihar region. During the year 1994-95 litchi was grown in a area of about 21.68 thousand hectares with a total production nearly 2.66 lakh tonnes. Bihar, thus, produces nearly 73.38 per cent of total litchi production of the country and occupies nearly 40 per cent of the area under litchi plantation in India. However, inspite of increase in litchi production, the litchi exported to the important cities of India from the state was only 7,702 MT and outside the country was only 30 MT during the year 1994-95. It was, therefore, pertinent not only to enquire into the growth rate of area, production and productivity of litchi in Bihar as well as Muzaffarpur district (which is the most important district of Bihar with respect to area and production) but also the export of litchi from the state.

Status of horticultural crops reveals that the mango contributes maximum area and production under total fruit crops grown in Bihar. The area and production of mango was 53.59 per cent and 53.08 per cent respectively. Banana attains second position in area and production. Litchi attains fourth position in the total fruit area and production. Inspite of alternate bearing in mango fruiting it covers 53.59 per cent of total fruit area in Bihar. It is only due to its wide range of soil and climatic adaptability, however, mango crop grown in less care and management and its market value is also very good.

The compound growth rates of area, production and productivity of litchi in Bihar for the period during 1980-81 to 1994-95 were estimated to be 2.8, 4.9 and 1.7 per cent respectively, which were positive and statistically significant. The growth rates of area, production and productivity during different sub-periods were also worked out and were observed to be positive and statistically significant.

The compound growth rates of area, production and productivity of litchi in the district of Muzaffarpur during the period 1980-81 to 1994-95 were observed to be 4.9, 6.6 and 1.7 per cent respectively, which were also positive and statistically significant. The growth rate of area production and productivity of litchi in the district of Muzaffarpur in different sub-periods were also positive and significant as it was for the State of Bihar. However, it was also observed that during 1990-91 to 1994-95 growth rate of area and production of Bihar (6.4% and 6.7%) was higher as compared to district of Muzaffarpur (4.5% and 6.4%). This indicated that the rate of growth of area and production of litchi in other litchi growing district of Bihar after nineties was faster as compared to district of Muzaffarpur, probably because litchi was already cultivated on large scale and opportunity for further expansion was comparatively less in Muzaffarpur than other litchi growing districts.

GUJARAT

In Gujarat with a total area of 446 hundred hectares under fruits in 1971-72, mango with 45.2 per cent of total area claimed a lion's share, which was followed by banana with 28.5 per cent and ber with 6.5 per cent, sourlime with 6.1 per cent. The area under chiku accounted for as low as 2.7 per cent of total area under fruits in the State.

In 1981-82, total area under fruits increased to 699 hundred hectares and accounting for 56.7 per cent increase during 1971-72 to 1981-82. The share of area under mango to total area under fruits was 37.9 per cent and that of banana was 30.6 per cent and chiku 6.0 per cent during 1981-82. As compared to 1971-72, the area under chiku, sourlime and other fruits increased in 1981-82.

In 1992-93, total area under fruits increased to 1065 hundred hectares, which accounted an increase of 138.8 per cent over 1971-72. The share of area under mango to total area under fruits was 31.9 per cent and that of banana was 26.9 per cent and chiku 5.6 per cent during 1992-93. With compared to 1981-82 the area under ber and other fruits increased in 1992-93.

In 1994-95 total area under fruits increased to 1290 hundred hectares, constituting an increase of 189.2 per cent over 1971-72. The share of area under mango to total area under fruits was 30.2 per cent followed by banana with 23.6 per cent, chiku 6.2 per cent and others 22.5 per cent. 'Others' category included fruits like pomegranate, detepalm, papaya, custardapple, jamun etc.

It revealed from the fact that on the whole area under total fruits increased by 189.2 per cent during 1994-95 over 1971-72. The share of chiku in total fruit area of the State increased from 2.7 per cent in 1971-72 to 6.2 per cent in 1994-95.

Area under total vegetables increased from 397 hundred hectares (0.4 % GCA) in 1971-72 to 1042 hundred hectares (0.9 % of GCA) in 1992.93.

Area under chiku in the State increased from 60 hundred hectares, constituting 5.4 per cent of area under total fruits of the State in1992-93 to 80 hundred hectares, constituting 6.2 per cent of area under total fruits of the production of chiku crop increased from 720 hundred MT in 1992-93 to 975 hundred MT in 1994-95. The yield was 12 MT per hectare during the period.

The area under chiku in Gujrat is projected to increase from 10750 hectares in quinquennium 1992-97 to 14000 hectares in quinquennium 2002-2007.

HARYANA

Unlike some of the other states of India, where flower cultivation for domestic consumption has been practised for a number of years, floriculture is relatively new to Haryana. In response to growing demand of flowers and cut-flowers from both domestic consumers as well as for export purposes, commercial cultivation of flowers in the State has started gradually picking up. The Haryana State Department of Horticulture has identified the districts of Sonepat, Gurgaon, Karnal and Ambala, in particular, for introduction of commercial floriculture in the State. As part of the promotional efforts, the Department of Horticulture has been organising demonstrations, arranging training programmes, and making available quality planting material of exotic varieties from different sources. Gladioli, roses, marigold, tuberoses, carnation and chrysanthemum have in particular been selected for commercial propagation and adoption.

As a result of these efforts and the locational advantage of Haryana, due to its proximity to Delhi which offers Haryana a ready and huge market for its produce, the area under floriculture has increased from just 50 hectares in 1990-91 to more than 1800 hectares in 1995-96.

While various species of flowers are cultivated in Haryana, marigold alone accounts for the largest share of flower cultivation about 63 per cent of the total area under flowers in Haryana. In addition about 8 per cent of flower area is occupied by each of the flowers- Gladioli, Roses, Chrysanthemum, and Tuberose. Over the years, the area under all types of flowers has increased, though to varying extent. The area under marigold has more than doubled- from about 550 hectares in 1991-92 to 1140 hectares in 1995-96. During the last four years however the area under carnation has remained almost constant at 10 hectares. Primarily as a result of increased area under different flowers, their production has increased significantly. Thus during the period 1991-92 to 1995-96, the annual production of gladiolus increased from 20 lakh to 320 lakh, that of roses from 40 lakhs to 125 lakh cut flowers, and that of chrysanthemum from 200 tons to 1200 tons. During the same period the production of marigold however more than doubled - from 12500 tons in 1991-92 to 28000 tons in 1995-96.

HIMACHAL PRADESH

In India the plantation of apples started with the introduction of apples mostly of sour varieties by the Christian Missionaries and the British Officers in the middle of 19th Century. The credit of cultivation of commercial apples goes to Samuel Evans Stokes a resident of Philadelphia who brought the delicious group of cultivars from United States of America in the beginning of 20th Century.

Presently more than 30 per cent of apple produced in the world is in China followed by U.S.A. contributing 11 per cent in 1995. In India the total area under apples was 1,30,000 hectares during 1978 and was increased to 2,10,734 hectares in 1995. During the same period the total production increased from 7,42,000 tonnes to 11,83,142 tonnes. The share of India in the world apple production is estimated to be 2.53 per cent and ranks ninth position in the world.

The state wise figure reveals that Jammu & Kashmir, Himachal Pradesh and U.P. hills account for more than 97 per cent of total area under apples in India and out of this Himachal Pradesh share was 35.81 per cent during 1994-95. Out of the total area and production of various fruits grown in Himachal Pradesh apple alone accounts for about 40 per cent and 86 per cent share respectively of all fruits grown in the State.

After the attainment of full statehood, Himachal Pradesh has achieved a very impressive progress in the production of horticultural crops (specially fruit and vegetables) because of planned efforts made by state government for the development of these crops. The farmers of the state also realised and accepted these crops in their farming system in place of traditional (subsistence) crops. Realising the potential of fruit crops in some of the areas of the state the field crops have been completely substituted with fruit crops.

During last two decades the area under fruit crops has recorded a compound growth rate of 3.18 per cent per annum and during the same period the production was increased at a compound growth rate of 3.64 per cent per annum. Among all the fruit crops temperate fruits have shown highest growth (more than 8 per cent per annum) in area under these crops. Though the area under all fruits have increased significantly changed in favour of new crops. Out of total fruits produced in the state, apple alone accounts largest share (more than 85 per cent) registering a compound growth rate of 3.60 per cent per annum.

In Himachal Pradesh fresh vegetables specially off-season vegetables accounts significant share in the horticultural crops of the State. During the last two decades the area under vegetables excluding potato have increased at a compound growth rate of 6.05 per cent per annum and production at the rate of 11.84 per cent per annum. Like fruit crops the adoption level of vegetables in the farming system of some of the regions is quite significant. The important valleys like Saproon in Solan, Sainj in Shimla and Nagwain in Mandi and Bajaura in Kullu districts, the adoption level of vegetable cultivation (specially off-season) is on commercial scale. In a recent study conducted in Saproon valley of Solan district the average returns from vegetable cultivation is estimated to be Rs.87 thousand per annum per farm.

The share of Shimla and Kullu districts in total fruit area in the stage have declined from 29.9 and 16.6 per cent to 18.9 and 11.8 per cent respectively during the last two decades. However in foot hill regions of the state some sub-tropical fruits becoming popular. In these areas the change in share of fruit crops have been increased significantly, specially in Kangra district where the share of fruit crops have increased from 13 per cent in 1975-76 to 21 per cent in 1995-96. Similarly in other foot hill districts like Una, Hamirpur, Bilaspur, the share of fruit crops in the cropping pattern of the region is increasing.

The overall scenario of the horticulture crops in the state suggest that some new crops are becoming popular and grown on commercial scale by harnessing the potential of for growing of these crops in various agro-climatic regions of the state.

KERALA

Kerala is a land of backwaters, forty four rivers cut across the State with their innumerable tributaries and branches. The backwaters are attractive and economically valuable feature of the land. The State with its high population presents complex problems in the sphere of food, employment and housing. The State is 50 per cent short of food due to historical and climatic conditions as the State has adopted commercial agriculture more than food crops. Rice is the staple food of the people. Out of a gross cropped area of 30.21 lakh hectares in 1992-93, the share of food crops (cereals, millets, pulses and tapicaco) was only 23 percent. Kerala has a unique cropping pattern, it accounts for 92 per cent of India's rubber, 70 percent of coconut, 60 per cent of tapiaco and almost 100 per cent of lemon gross oil. The state is a single largest producer of a number of other crops like banana and ginger besides tea and coffee in abundance. Coir and cashew are two of the largest traditional industries in the kerala state.

Cashew is one of the most important export oriented horticultural crops in kerala which is a native of Brazil was introduced to India in the sixteenth century by the early portuguese settlers. It was originally grown for the purpose of checking soil erosion owing to its quality of resistance to a wide range of vagaries of nature. Due to conditions peculiarly suitable for its growth there was rapid spread along the entire West Coast and South East Coast of india and over years it has established itself in this country as an important commercial crop.

MAHARASTRA

In terms of fruit and vegetable production Maharastra is considered to be the most important state of the country. During the last ten years period, there has been significant increase in the area and production of horticultural crops in the state. The area under horticultural crops in Maharastra grew from 3.76 lack hectare during 1981-82 to 6.40 lack hectare in 1991-92 registering a nearly two fold increase in the area. It is to be noted that this state leads the country in production of grapes, bananas, orange and onions. Grape is already established as an important commercial crops in Maharastra. Although the cultivation is mainly is concentrated in three districts of Nasik, Sangli and Solapur. Large number of farmers in the neighbouring districts like Pune, Ahmedanagar and Satara are switching over to grape cultivation. Infact, grape cultivation is chiefly confined to deacon platen in western Maharastra because of the congenial agro-climatic conditions prevailing in this region. About 600 hectares are added to grape acreage in Maharastra every year. Nasik district of Maharastra is the largest producer of grapes in to the country. It accounts for a production of about 1,80,000 tonnes of grape per annum. The yield of grapes in Nasik has risen from 12 tonnes per hectares a few years age to 22 tonnes per hectare. Even harvesting of grapes is done twice as against once in most of the countries.

Another important horticultural crop grown in Maharastra is onion. Most of the onion produced in India comes from this state. This state accounts for 20 per cent of the area and 25 per cent of the total output of onion in India. Nasik not only producing grapes, this district also accounts for the bulk of the onion production in the state. In fact, Nasik accounts for the largest share in the production of onion in India. Onion is also grown in Pimpalgaon, Manmad, Yeola, Saikheda, Chandwad and Satana all around Nasik.

SIKKIM

The cultivation of orchid in the state of Sikkim has been increasing at a rapid rate due to more or less better return and external demand of these plants. Therefore, the need for assured supply of quality orchid to capture and ensure the foreign market niche in a competitive global market environment, is being deeply felt. This requires commitment at production and product development stages. A synchronised production strategy at the micro-level and an export policy at the macro-level are the calls for the day.

In recent years different schemes and programmes have been implemented by the Government of Sikkim in consonance with the strategies formulated for the Eighth Plan Period. It entails development of the horticultural sector more vigorously with special emphasis on cash crops. The development of floriculture and off-seasons vegetables have also received over riding priority. However, some improvement has been found in the production of fruits and vegetables in this state during the last decades. The macro level analysis regarding the trend in production of horticultural crops in Sikkim shows that the production of total fruits has increased from 4,70 thousand tonnes in 1975-76 to 12.00 thousand tonnes in 1995-96. The production of total vegetables in the state was estimated to be 28.00 thousand tonnes and that of potato - the major vegetables - at 24.00 thousand tonnes in 1995-96. The production of large cardamom and ginger was around 3.60 to 29.00 thousand tonnes respectively during the same period. The percentage growth rates of area of orange/citrus fruits and other fruits during the period from 1975-76 to 1995-96 have increased by 471.43 per cent and 173.33 per cent respectively. The percentage growth rate of area and production of large cardamom during this period were 235.00 per cent and 156.52 per cent respectively. Another important horticultural commodity is ginger which occupies 8.98 per cent of the total area under horticultural crops. The area and production of ginger in Sikkim have registered the rates of increase at 900 per cent and 1450 per cent during the same period.

TRIPURA

There has no systematic study on area and production of horticultural crops in the region. The North Eastern Council estimated that about 1.93 lakh hectares of land are occupied by various fruit crops of which 36 thousand hectares are under pineapple. The productivity of horticultural crops are not encouraging, yet, the estimated marketable surplus in the region is highest for pineapple (82 per cent) followed by mandarin orange(76 per cent) and banana (64 per cent). So far as allocations of area under pineapple is concerned Assam ranked first followed by Meghalaya, Manipur and Tripura. The compound growth rate of area and production during 1980-81 to 1991-92 has been worked out at 5.37 and 5.59 per cent respectively.

Considering the quantum of production and marketable surplus (88%) in the sample it is identified as an export oriented horticultural crop of Tripura. The official level information indicated that a considerable improvement of export of horticultural crops has been registered in 1996 over 1995. The quantum of export of pineapple to Bangladesh was 10 MT. in 1995 which has increased to 134.25 MT. in 1996 and so is the case with citrus and jak fruits. But the quantum of export to total marketable surplus has been considered as nominal. It is gathered that a considerable amount of pineapple, jackfruits, citrus etc. are exported to Bangladesh through some illegal means which has got no record at official level.

The sample farmers owned 207.61 hectares of land of which 127.73 ha. (61.52%) are under pineapple, 27.20 ha. (13.10%) under field crops. Sugarcane and Rubber plantation is the recent introduction in the area. In the sample the percentages of marginal, small, semi-medium, medium and large farmers are 16,24,36, 20 and 4 respectively.

UTTAR PRADESH

The major horticultural crops in Uttar Pradesh were categorised into fruits, vegetable and spices. During 1993-94 the total area covered by the fruit crops was 482.79 thousand hectares of which the maximum i.e., 258.43 thousand hectares covered by the mango crop only whereas guava and musk melon are the other major horticultural crops. The area covered by other fruit crops was recorded 52.21 thousand hectare. In case of Mango, district Saharanpur, Lucknow and Sultanpur are the major growing districts of Utter Pradesh.

In case of vegetables 602.37 thousand hectares of area was under these crops out of which 378.30 thousand hectares was under potato crop during 1993-94. Sweet potato and onion are the another vegetable crops grown in U.P. Among the regions of plain the area under mango was found to be highest in Western region followed by eastern and central region. The production of mango was also observed on an increasing order. Regarding area under potato it was in increasing order till the year of 1992-93 but thereafter the area under said crop started decreasing. During 1991-92 the area under potato was 368.94 thousand hectares which decreased to 360.61 during 1994-95. The production and productivity of potato seems to be increasing in very slow pace.

WEST BENGAL

The cultivation of orchid in the state of West Bengal has been increasing at a rapid rate due to more or less better return and external demand of these plants. Therefore, the need for assured supply of quality orchid to capture and ensure the foreign market niche in a competitive global market environment, is being deeply felt. This requires commitment at production and product development stages. A synchronized production strategy at the micro-level and export policy at the macro-level is the call for the day.

The macro level analysis regarding the trend in area and production of horticultural crops in West Bengal shows that in spite of the wide adaptability of the fruit crops like mango, pineapple and banana, these are mainly confined to three/four districts of the state. The compound growth rates of area during the period from 1990-91 to 1992-93 under mango, pine apple and banana show that these are increased by 1.89, 2.57 and 12.31 per cent respectively in the state, while the area under vegetables in different seasons i.e. summer and rainy vegetables are increased by 17.29 and 113.08 per cent respectively in the state during the same period. However, the compound growth rate of area under winter vegetables is found to be -8.79 per cent during the same period. The production of these horticultural crops in the state also demonstrate an increasing trend during the period under review except mango and winter vegetables which are estimated to be negative.

ROLE OF APEDA IN PROMOTING AGRICULTURAL EXPORTS

The Agricultural and Processed Food Products Export Development Authority (APEDA) was established by the Government of India under the Agricultural and Processed Food Products Export Development Authority Act passed by the Parliament in December, 1985. The Act (2 of 1986) came into effect from 13th February, 1986 by a notification issued in the Gazette of India: Extraordinary: Part-II [Sec. 3(ii): 13.2.1986). The Authority replaced the Processed Food Export Promotion Council (PFEPC).

Functions of APEDA

In accordance with the Agricultural and Processed Food Products Export Development Authority Act, 1985, (2 of 1986) the following functions have been assigned to the Authority.

Development of industries relating to the scheduled products for export by way of providing financial assistance or otherwise for undertaking surveys and feasibility studies, participation in enquiry capital through joint ventures and other reliefs and subsidy schemes.

Registration of persons as exporters of the scheduled products on payment of such fees as may be prescribed.

Fixing of standards and specifications for the scheduled products for the purpose of exports.

Carrying out inspection of meat and meat products in slaughter houses, processing plants, storage premises, conveyances or other places where such products are kept or handled for the purpose of ensuring the quality of such products.

Improving of packaging of the scheduled products.

Improving of marketing of the scheduled products outside India.

Promotion of export oriented production and development of the Scheduled products.

Collection of statistics from the owners of factories or establishments engaged in the production, processing, packaging, marketing or export of the scheduled products or from such other persons as may be prescribed on any matter relating to the scheduled products and publication of the statistics so collected or of any portions thereof or extracts therefrom.

Training in various aspects of the industries connected with the scheduled products.

Products monitored by APEDA

APEDA is mandated with the responsibility of export promotion and development of the following scheduled products:

Fruits, Vegetables and their Products.

Meat and Meat Products.

Poultry and Poultry Products.

Dairy Products.

Confectionery, Biscuits and Bakery Products.

Honey, Jaggery and Sugar Products.

Cocoa and its products, chocolates of all kinds.

Alcoholic and Non-Alcoholic Beverages.

Cereal and Cereal Products.

Groundnut, Peanut and Walnut.

Pickles, Papads and Chutneys.

Guar Gum.

Floriculture and Floriculture Products.

Herbal and Medicinal Plants.

In addition to this, APEDA has been entrusted with the responsibility to monitor import of sugar.

Trade promotion assistance

Marketing development assistance (MDA) scheme

It is under operation through the Department of Commerce to support the under mentioned activities:

Assist exporters for export promotion activities abroad.

Assist Export Promotion Councils (EPCs) to undertake export promotion activities for their product(s) and commodities.

Assist approved organizations/trade bodies in undertaking exclusive nonrecurring innovative activities connected with export promotion efforts for their members.

Assist Focus export promotion programmes in specific regions abroad like FOCUS

(LAC), Focus (Africa), Focus (CIS) and Focus (ASEAN + 2) programmes.

Residual essential activities connected with marketing promotion efforts abroad.

Assistance to individual exporters for export promotion activities abroad or Participation in EPC etc. led BSMs/Trade Fairs/ Exhibitions:

Sr. No.	Area /Sector	No. of Visit	Maximum Financial Ceiling per event
1	Focus LAC	1	Rs 2,50,000
2	Focus Africa (Including WANA)	1	Rs 2,00,000
3	Focus CIS	1	Rs 2,00,000
4	Focus ASEAN + 2	1	Rs 2,00,000
5	General Areas	1	Rs 1,50,000
	Total	5	

LAC: Latin American Countries : 26, ASEAN: Association of Southeast Asian Nations : 8+2 CIS: Commonwealth of Independent States (former Soviet Republics)

The participation of individual companies in the above activities shall be subject to the following conditions: Any exporter having FOB value of Export up to 30 crores.

Schemes for market development

Sr No.	Components	Pattern of Assistance
1	Assistance to exporters for use of packaging material as per standards and specifications developed or adopted by APEDA	25% of the total cost of packaging material (including inner packaging materials viwz. punnets, sleeves, rubber band etc), subject to a ceiling of Rs 5 lakh
2	Assistance for conducting feasibility studies etc.	50% of the total cost subject to a ceiling of Rs 10 lakh
3	Conducting surveys, feasibility studies etc. for the common benefit of a number of exporters who may be the members of the Associations	Assistance will be in line with the guidelines of MDA Scheme
4	Brand publicity (Product specific Indian Brands) through advertisement in international print/ electronic media etc.	25% of the total cost subject to a ceiling of Rs 50 Lakh.Not to be extended to an exporter beyond 3 consecutive years.

Scheme for infrastructure development

Sr No.	Components	Pattern of Assistance
1	Establishment of common infrastructure facil- ities by APEDA or any other Government or Public Sector	90% grant-in-aid by APEDA and 10% from other govt /pub- lic sector
2	Purchase of specialised transport units for ani- mal products, horticulture and floriculture sector	40% of the cost subject to a ceiling of Rs 7.5 lakh
3	Setting up of sheds for intermediate storage and grading / storage / cleaning operation of pro- duce.	40% of the cost of equipment subject to a ceiling of Rs 10.00 lakh
4	Setting up of mechanized handling facilities such as sorting, grading, washing, waxing, rip- ening, packaging & palletisation, pre-cooling, cold storage, fumigation, X-ray screening, hot water dip treatment, Water softening plant, etc.	40% of the cost of equipment subject to a ceiling of Rs 25.00 lakh
5	Setting up of integrated post harvest-handling system (pack houses with any two or more of the above facilities	40% of the cost subject to a ceiling of Rs 75.00 lakh
6	Setting up of cable cars (covering minimum of 50 ha of plantation) for banana and other crops	-do-
7	Setting up of vapor heat treatment, electro-nic beam processing or irradiation facilities	40% of the cost subject to a ceiling of Rs 50 lakh

Sr No.	Components	Pattern of Assistance
8	Setting up of environment control system e.g. pollution control, effluent treatment etc	40% of the cost subject to a ceiling of Rs 35 lakh
9	Setting up of specialized storage facilities such as high humidity cold storage deep freezers, controlled atmosphere (CA), etc for fresh and processed horticultural produce	40% of the cost subject to a ceiling of Rs 25 lakh

Schemes for quality development

Sr No.	Components	Pattern of Assistance
1	Assistance for installing quality testing equipments	50% of the cost subject to a ceiling of Rs 25 lakh.
2	Installing quality management, quality assurance and quality control systems such as ISO series, HACCP, BRC, GAP etc. including consultancy and certification	50% of the cost subject to a ceiling of Rs 5 lakh.
3	Testing of water, soil, residues of pesticide, veterinary drugs, toxins, heavy metal contaminants in fruits and vegetables, processed products, floriculture, animal products.	50% of the total cost subject to a ceiling of Rs 5000/- per sample where residue monitoring is proposed by APEDA.

4	Adoption of global standards for traceability registration for Global Company Prefix Number (GCP), Electronic Product Code (EPC/RFID)	25% of the fee subject to a ceiling of Rs 1 lakh.

Schemes for capacity building

Sr No.	Components	Pattern of Assistance
1	Upgradation of APEDA recognized labs for export testing	50% of the cost for private labs; 75% for State Govt. /University lab; and 100% for the Central Government labs; The above is subject to a ceiling of Rs 75 lakh.
2	Assistance for technical skill upgradation of personnel of exporters in India for exportdevelopment	Rs 20,000 per personnel per residential training (not more than 5 personnel from single organization per year) in arecognized training institutes. Assistance will be limited to an average cost of Rs 5000 per day per personnel 100% for APEDA representatives.
3	Assistance for organizing seminars/ group activities including study tour within the country and for bringing out information literature	50% of the total cost subject to a ceilingof Rs 1 lakh for national seminar and Rs 2lakh for international seminar.
4	Assistance to carry out relevant applied Research & Development activities for export enhancement	100%byAPEDAthroughICARinstitutions or State Agriculture Universities. In case of assistance to exporters associations in cooperative/private sector up to 50% of the total cost subject to a ceiling of Rs 50 lakh per project.

Scheme for transport assistance

The norms for TA calculation are as follows:

TA Norms for Exports by Air
Least of:
10% of FOB Value
25% of Freight
Specific rate (Rs per kg)

TA Norms for Exports by Sea	
For eligible products exported in non-reefer containers	For eligible products exported in reefer containers
Least of:	Least of:
10% of FOB value	10% of FOB value
25% of Basic Ocean Freight	33% of Basic Ocean Freight (inclusive of inland freight in reefer containers & basic ocean freight) (see note below)
Specific rate (Rs per kg)	Specific rate (Rs. per kg)

* Note: Inland freight by Reefer container (from factory to port) will be included in the freight norm used for TA calculations subject to the following conditions:

The reefer container is factory stuffed (i.e. inland movement from factory to port is by reefer container truck).

Copies of the excise and customs bonded seal (for inland movement) are included with the TA claim.

The total quantum of TA does not exceed 50% of the ocean freight.

Eligible items and permissible destinations for air / sea transport assistance are listed below:

A. By air

Eligible Products/Item	Destinations	Minimum FOB Value (Rs/ Kg) for eligibility
Floriculture		
Fresh Cut Flowers/ Bouquets	All Destinations except neighboring countries	100
Live Plants and Bulbs		100
Tissue Cultured Plants		2000
Fresh Fruits		
Fresh Fruits excluding Mango and Grapes	All Destinations except neighboring countries	30
Grapes	All destinations except neighboring countries, EU and Middle East Countries	35
Mangoes		30

Fresh Vegetables		
Fresh Vegetables (including organically grown vegetables and mixed vegetables excluding fresh Mushrooms)	All Destinations exceptneighboring countries	30
Fresh Mushrooms		35

A. By air

Eligible Products/Item	Destinations	Minimum FOB Value (Rs/Kg.) for eligibility
Fresh/ Chilled Fruits		
Fresh Fruits excluding Mango and Grapes	All Destinations except neighboring countries	30

Eligible Products/Item	Destinations	Minimum FOB Value (Rs/Kg.) for eligibility
Grapes	All destinations except neighboring countries, EU and Middle East Countries	35
Mangoes		30
Fresh/ Chilled Vegetables		
Fresh/ Chilled Vegetables (including organically grown vegetables and mixed vegetables)	All Destinations except neighboring countries	30
Processed Vegetables		
Preserved Vegetables (Acetic Acid/ Vinegar/Brine) (including Gherkins prepared and preserved, Onions inAcetic Acid and Potato Flakes/ Powder but excluding Pulses)		35
Dehydrated Onion & Garlic (Flakes/Powder	All Destinations exceptneighboring countries.	70
Pickles/ Chutneys (Fruits or Vegetables or mixture of Fruits and Vegetables)		60
Mushrooms Prepared/ Preserved		50
All processed Vegetables in Frozen/ IQF form		60

Processed Fruits		
All Processed Fruit Preparations (Excluding squashes, juices and pulps)	All Destinations except neighboring countries.	50
Fruit Squash		60
Fruit Juices		60
Fruit Pulps		60
Floriculture		
Fresh Cut Flowers	All Destinations except neighboring countries.	100
Dried Flower		75

Transport assistance (by Air)

Specific TA rates for exports of eligible items by Air

Destination	Specific TA Rate (Rs Per kg)	
	For all eligible products/ items except Fresh Cut Flowers	Fresh Cut Flowers
Europe	17	30
Africa	15	30
Middle East	7	10
CIS Countries	7	12
South East Asia	7	12

East & North Asia (including China Japan)	18	30
North America (USA, Canada & Mexico)	25	45
Central America & Caribbean	25	45
South America	25	45
Australia & New Zealand	18	34

Transport assistance (by Sea)

Specific TA rates for exports of eligible terms by sea

Destination	TA Rate (Rs Per kg)	
	Reefer Container	Non-Reefer Container
Europe	6.0	1.5
Africa	6.0	1.5
Middle East	3.0	1.0
CIS Countries	6.0	1.25
South East Asia	3.0	1.0
Far East & North Asia (including China Japan)	6.0	1.25
North America (USA, Canada & Mexico)	8.0	2.5
Central America & Caribbean	8.0	2.5
South America	8.0	2.5
Australia & New Zealand	6.0	1.25

Region wise % basic ocean freight

SR. NO.	REGION	% BASIC OCEAN FREIGHT
1	Africa	70.00
2	Australia	70.00
3	Europe	61.00
4	Middle East Asia	73.00
5	North America	74.00
6	North Asia	63.00

7	South America	78.00
8	South Asia	71.00

Online services

I-TRACK SYSTEM
Register as Member
Exporte'r Login
Apeda's Internal User
TRACEABILITY
HORTINET : Grape / Pomegranate /Mango / Vegetables
PEANUTNET
TRACENET
MEATNET

Challenges in export of fruits and vegetables

Although India is a leading fruit producer in the world, the fresh fruit export from India is very small owing t o a number of constraints. These constraints relates to production practices, post harvest technologies, issues related to supply chain, market access and non- tariff restrictions and governmental policies.

Constraints for Export

India continues to be absent or at best a marginal player in most of the leading markets for its export of fresh fruits. Indian players have not succeeded in establishing direct linkages with buyers/consumers in importing countries, as a result of which a large proportion of exports are being further processed and re-exported by other countries. Some of the major concerns for promoting the export relate to-

Lack of exportable varieties (high fiber content, inappropriate appearance and texture and large size of stone)

Lack of post-harvest treatment facilities such as of vapour heat treatment

Lack of pack houses from farm to port

High cost of obtaining certification for exports like Eurep-Gap

The issues can be categorized into "Supply Chain Issues" which are inherent to the domestic supply chain of food products and "Market Access Issues" which comprise of various parameters and factors driven by the requirements of the target countries.

Supply Chain Issues

Uneconomic scale of operation

Lack of consistency in supply and quality

Lack of cost competitiveness due to statutory changes, intermediation and wastages losses

Inadequate and inappropriate storage and distribution infrastructure

Lack of technical support for the agro-industrial sector

Market Access Issues

Non-Tariff Barriers: Non-tariff barriers can take various forms. Broadly these can be categorized as under:

Import Policy Barriers

Standards, Testing, Labelling and Certification requirements

Anti-dumping and Countervailing Measures

Export Subsidies and Domestic Support

Government procurement

Short product life cycle

Lack of brand image

Most exports from India lack scale- for example the largest fresh produce exporter records annual sales of about Rs 500 million. The low volume translates into lack of economics in operation and mares exports uncompetitive. Hence, exporters are not able to establish themselves as long term players in the export market, and rely heavily on opportunistic businesses. These factors cumulatively translate into low investments in upgrading skill sets, product innovation, quality improvement and brand building.

Multiple Safety standards

The emergence of private food safety and quality standards mainly in developed countries is now a well-established fact. These standards operate alongside regulatory systems but in terms of market access and access to the shelves of the leading supermarkets in the rich countries, it become almost mandatory. With these standards becoming a global phenomenon, countries in the developing world face increasing constraints in exporting their food products to markets in the developed countries.

Technological Constraints

The major technology related constraints contributing to low productivity of horticultural crops and inferior quality of produce are:

Vast majority of holdings are small and un-irrigated

Large tracts of low and unproductive plantations needing replacement/ rejuvenation.

Low productivity of crops due to inferior genetic stocks and poor management.

Inadequate supply of quality' planting materials of improved varieties o High incidence of pests and diseases.

Heavy post harvest losses and low utilization in processing sector.

As per available literature, several studies have been conducted by various export institutions like APEDA ICAR, NHB, NABARD regarding constraints in export of fruits and vegetables. They have highlighted numerous problems. A few of them are listed below-

Complex supply chain is one of the main barriers in export of fruits and vegetables. In markets a number of small producers, commission agents are operating. Therefore, for a bulk purchase of lot, these numbers of functionaries are to be dealt, causing inconvenience and more post-harvest losses.

Lack of continuous and adequate supply. Weather fluctuations, causing interrupted supply of fruits and vegetables in the market. Sometimes, weather also delayed arrivals, keeping prices on the boil for certain period.

Inconsistent supply of exportable quality.

Lack of forward/backward linkage between growers/suppliers and buyers/ exporters which affects the market potential.

Highly perishable (short storage life and susceptible to diseases) resulting in high post-harvest losses.

High freight cost and inadequate transport facilities.

The disparity in wholesale and retail prices.

Complex Procedures for Sampling/Product Testing

High Cost of Certification

Frequent and sudden impose of ban by importing country

Commercial constraints

According to the USDA economic report, the commercial constraints on fruits and vegetables include:

Trade barriers: Natural trade barriers include high transportation costs to distant markets, and artificial barriers include legal measures such as protectionist policies. Liberalization of trade through international agreements has been instrumental in relaxing many legal trade barriers by reducing tariffs and by harmonizing the technical barriers to trade.

Scientific phyto-sanitary requirements: Importing countries set the standards that potential trade partners must meet in order to protect human health or prevent the spread of pests and diseases. For instance, Japanese imports of US apples are limited to Red and Golden Delicious apples from Washington and Oregon. The Japanese, who are mainly concerned with the spread of fire blight, impose rigorous and costly import requirements on the US apple shippers. The apples must be subjected to a cold treatment and fumigation with methyl bromide before shipment to Japan, and three inspections of US apple orchards during the production stage. Infestation by fruit flies (Tephritidae: Diptera), common in the tropics, is a major constraint to the production and export of tropical fruits.

Technological innovations: Countries can increase their competitiveness and world market shares by providing higher quality products and promoting lower prices through technological innovations.

Table 1.3 represents the problems faced by exporters of fruits and vegetables from Gujarat. According to Researcher, different problems were identified in Gujarat through random survey method and for getting mean score Garret Ranking technique was employed.

Table 1.3 Problems faced by exporters of fruits and vegetables from Gujarat

Sr. No.	Exporters perceives problems for export of fruits & Vegetables	Mean Score	Rank
1	Difficulties in certification	4.0	1
2	Difficulties in custom clearance	3.9	2
3	Problems in quoting price with fluctuating exchange rates	3.9	3
4	Sanitary and Phytosanitary (SPS) measures	3.8	4
5	Price fluctuation of the commodity supply at home country	3.7	5
6	Problem finding reliable foreign distributor	3.6	6

7	Competition from firms in foreign markets	3.5	7
8	Difficulty in collecting payment from customers abroad	3.5	7
9	Lack of capital to finance expansion into foreign markets	3.4	8
10	Lack of capacity dedicated to continuing supply of exports	3.3	9
11	Confusing foreign import regulations	3.3	9
12	Informational barrier	3.3	9
13	Lack of facilities like pre-cooling, cold-storage, pack house, etc.	3.1	10
14	Labelling and packaging requirement	3.1	10
15	Transportation cost and duration	3.1	10

The results revealed that the exporters of fruits and vegetables from Gujarat were facing several problems among which difficulties in certification ranked first followed by difficulties in custom clearance, problems in quoting price with fluctuating exchange rates, sanitary and phyto-sanitary (SPS) measures, price fluctuation of the commodity supply at home country, problem finding reliable foreign distributor, competition from firm in foreign markets, difficulty in collecting payment from customers abroad etc.

EXPORTS AND IMPORTS

Export of Agricultural Products

Exports of agri-horti products are dependant on factors such as domestic production and consumption, exportable surpluses, consumer preferences, varieties traded, quality, domestic and international prices and availability of infrastructure facilities for storage, post harvest handling, etc. Having regard to the social and economic importance of the agricultural sector, the export strategy of the Government is based on the premise that foreign earnings from this sector should be enhanced thereby leading to higher income to farmers, taking care to make agricultural products available at reasonable prices to the domestic consumers.

The exports of agricultural products which were valued at Rs.77600 million at the beginning of the VIII Plan have virtually doubled to Rs.1,50,000 million by 1996-97, the end of VIII Plan. At the end of the financial year 1998-99, the exports of Agricultural products have further risen to Rs. 1,60,000 million (approx).

The agricultural products exported can be divided into three broad categories, ie., exports of raw products, semi processed products and export specific value added products. The raw products exported are of low value but high volume. Semi-processed products exported are of intermediate value and limited volume; while export specific and value added products are of high value and low volume.

Export of Horticultural Commodities

Among horticultural commodities, India is exporting fresh fruits, vegetables, processed products of fruits and vegetables, cut & dried flowers, medicinal and aromatic plants, seeds, spices, cashew kernels and their products, tea and coffee. The total value of export of these commodities increased from Rs.29722.78 million in 1991-92 to Rs.90315.20 million in 1998-99.

Table 1.4 Export of horticultural products from India

Products	1991-92		1998-99	
	Quantity ('000 tonnes)	Value (Rs. in million)	Quantity ('000 tonnes)	Value (Rs. in million)
Fruits & Vegetables	493.61	2934.50	387.43	5360.20
Processed fruits & vegetables	79.90	1777.80	238.60	7056.80
Fruits & Vegetable seeds	8.42	160.90	6.06	663.30
Floriculture	NA	148.00	18.72	966.00
Arecanut/ Cocoa	0.21	111.40	20.59	140.30
Coconut	31.00	745.50	56.00	3021.00
Cashew & its Products	47.74	4422.40	75.02	16099.00
Spices	142.10	3809.68	231.39	17580.20
Tea	216.45	12122.60	205.86	21918.40
Coffee	111.45	3490.00	211.62	17510.00
TOTAL	1130.88	29722.78	1451.29	90315.20

Horticulture products account for more than 54.55% (1998-99) of the total value of exports of agricultural commodities from India. Of these, export of tea leads all horticultural commodities followed by spices, coffee, cashew and processed fruits and vegetables. The status of export of individual commodities / groups of commodities is given in this chapter.

FRESH FRUITS & VEGETABLES

There has been significant increase in the export of fresh fruits and vegetables during the past few years. Exports increased from Rs.2934.5 million in 1991-92 to Rs. 5360.2 million in 1998-99. Among fruits mango was the main fruit exported. Export of fresh mango had commenced as early as 1925. Among vegetables onion has been the major crop exported from India. The Government of India has established Agricultural and Processed Food Export Development Authority (APEDA) under the Ministry of Commerce to promote fruits and vegetables export. The APEDA in turn has initiated a programme for an integrated training of horticulture producers for some identified fruits such as grape, mango, litchi, Kinnow in the selected regions. Farmers have been provided training in integrated post harvest management practices for better handling of the produce to ensure export of quality products. APEDA is also making efforts to enhance the shelf life of fruits such as mango, grape, litchi through use of controlled/ modified atmosphere storage and use of reefer containers so that they could be transported by sea freight and achieve higher competitive advantage. In order to improve quality of fruits and vegetables, pre-harvest manuals on certain fruits and vegetables have been prepared for dissemination to farmers and producers.

Fruits

During 1998-99, 96487.57 tonnes of fruits and nuts were exported (Table 1.5). The export earnings, which was Rs. 726.5 million during 1991-92 increased to Rs. 2384.48 million in 1998-99. Mango occupies a premier position among fruits valued at Rs. 791.37 million. Other fruits, which have attained significant position in export, are grape, walnut, citrus (Kinnow), banana and apple. Small quantities of a number of other fruits e.g. litchi, guava, custard apple, pineapple papaya and tamarind have also demand in the export market.

Mango varieties exported from India are Alphanso, Kesar, Dashahari, Banganapalli. Variety Thompson Seedless variety constitutes the bulk of exports of grape. Besides, Kinnow is other fruit exported, which has attained a sizeable status value-wise, however, the four fruits and nut exported are mango, grape, kinnow and walnut respectively.

Table 1.5 Export of major fruits from India (1998-99)

Product	Quantity (tonnes)	Value (Rs in million)
Apple	7442.12	100.24
Banana	8111.42	168.94
Citrus	12786.30	151.94
Grapes	11382.12	370.91
Guava	496.84	6.92
Mango	45407.59	791.37
Pineapple	244.68	1.68
Pomegranate	4239.15	89.61
Sapota	1049.45	13.69
Walnut (Whole + Kernel)	5327.9	689.18
Total	96487.57	2384.48

Vegetables

Fresh vegetable exports have been on the rise. During 1998-99, the total vegetable exports amounted to 238279.28 tonnes valued at Rs. 2104.23 million (Table 1.6). The major item exported is onion, worth 1760.47 million, with a share of 83.66% in vegetable exports. Other crops with significant exports include tomato (Rs.4.75 million) peas (Rs.1.63 million) and cucumber (Gherkin) (Rs.71.65 million). Mixed vegetable exports amounted to Rs.265.73 million, with the share of exports being 12.63%. Main vegetable exports from India are to South-east Asia and Middle East, except cucumber and gherkin which are exported to Europe and the U.S.A. High-value beans, peas, green chilli etc. have good scope for export.

Table 1.6 Export of major vegetables from India (1998-99)

Crop	Quantity (Tonnes)	Value (Million Rs.)
Cucumber (Gherkin)	5132.32	71.65
Mixed vegetables	16738.06	265.73
Onion	215693.61	1760.47
Peas	72.58	1.63
Tomato	642.71	4.75 2104.23
Total	238279.28	

Processed Fruits & Vegetables

Export of processed fruits and vegetables is another thrust area for increasing export of value added products. While processing industry in India has been dependent on small units without modern facilities a large number of multinational companies have entered the processed food industries in recent years with a thrust on exports.

Some of the new products which have been introduced in the market are tomato paste in bulk aseptic packs, freeze dried and Instant Quick Frozen fruits, vegetables and gherkins. The total export of processed fruits & vegetables from India is Rs. 7056.8 million. This consists 54.24 % of dried & preserved vegetables, 19.70 % of mango pulp,

15.30 % of other processed fruits & vegetables and 10.76 % of pickles & chutney. The APEDA in its quality upgradation programme has covered 26 Mango pulp processing units in Andhra Pradesh and Tamil Nadu for certification under the HACCP Quality Management tool.

Potato

Limited quantities of potato are exported from India as fresh potato, seed potato and frozen potato. Fresh potato exports during 1998-99 amounted to 20883.80 tonnes valued at 9.037 million. Frozen potato exports amounted to 1590.30 tonnes, valued at Rs. 1.104 million. Seed potato exports amounted to 431.64 tonnes, valued at Rs.0.469 million (Tables 1.7).

Major potato exports are to the neighboring countries of South Asia and South East Asia. Since most potatoes in India are harvested in spring, when fresh potatoes are not available in Europe, with a large production base and a sound export strategy, India has opportunities to exploit fresh potato as well as seed potato exports.

Table 1.7 Export of potatoes from India (1998-99)

Item	Quantity (tonnes)	Value (Rs. in million)
Potato fresh	20883.80	9.037
Frozen potato	1590.30	1.104
Potato seed	431.64	0.469 0.610
Total	22905.74	

Mushroom

Mushroom exports started in 1993-94 with 4811.478 tonnes. However, there has been a decline in mushroom export during 1998-99 with a total export of 3548.37 tonnes, mostly in dried/processed form (Table 1.8).

Table 1.8 Mushroom exports from India (1998-99)

Type of Mushroom	Quantity (tonnes)	Value (in million Rs.)
Dried	90.13	277.27
Processed	3458.24	139.27
Total	3548.37	416.54

Dried mushrooms were exported to 18 Countries, the major share of which went to France, Germany and Switzerland. The major importer of processed mushroom from India were USA, Israel, Denmark and Canada.

Flowers

Although India's share in the export market of flowers is still insignificant, it has registered a sharp increase from Rs.144.5 million in 1991-92 to Rs. 966.0 million in 1998-99. The cut flower export value have shown a tremendous increase during the period from Rs. 4 million to Rs. 253.0 million. This has come about with the establishment of a large number of export oriented cut flower units around Bangalore, Pune, Delhi and Hyderabad during the last five years. The major product has been rose being grown by more than 90 per cent of commercial units followed by tropical orchids (dendrobium). Limited exports are also taking place in carnation, geranium etc. The major destinations have been Europe (Germany, Holland & U.K.). The other markets importing Indian flowers are Japan, Australia, Russia and Singapore. A unit near Chennai is exporting tropical orchids. Other potential commodities for export are cut flowers, house plants, tissue culture material, dry flowers and hybrid seeds.

Dry flowers contribute a major share of the floriculture trade. Flower crops like dahlia, bell cup, marigold, jute flower, wood roses, wild lilly, helicrysum, lotus pods which can be easily processed and preserved as dry flowers hold enormous potential as these are becoming popular due to their non-perishability and are being exported from India.

Export of dry flowers and floral products has been going on since 1985. Indian flower trade has crossed 10,000 tonnes with major markets being US,

Israel, Hongkong, Japan, Singapore and West European countries. U.K. has been the largest importer of dried flowers from India, ahead of Germany, Italy, the Netherlands and Spain. The dry flower units are concentrated in places like Tuticorn in Tamil Nadu and Kolkata (Calcutta).

Medicinal and Aromatic Plants

Demand for medicinal and aromatic plants has been increasing progressively with the increase in the number of multinationals establishing consumer oriented cosmetics and pharmaceutical units. Currently, most of these herbs are grown in large quantities and marketed by France, U.K., Canada, Turkey and U.S.A. The foreign exchange earned by these countries through the export of these medicinal and aromatic plants and their products is quite substantial.

The potential for foreign exchange earning by India from the exports of medicinal and aromatic plants is estimated to be over U.S. $ 3000 million per annum. The exports of these plants and their products have a tremendous potential particularly to advanced countries like U.S.A., Japan and Europe. The demand for such plants is increasing both in developing and developed countries. The international market of medicinal plants related trade is estimated at US $ 60 billion per year having a growth rate of 7 per cent per annum. The annual exports of the derivatives of medicinal aromatic plants are to the tune of Rs. 600-700 million. A list of medicinal plants finding place in the international trade is given in Table 1.9.

Table 1.9 List of Indian medicinal plants entering into international trade

Sl.No.	Name of Medicinal Plant	Sl. No.	Name of Medicinal Plant
1.	*Cycas circinalis*	8.	*E. epidendracea*
2.	*Dendrobium candidum*	9.	*E. herbaceae*
3.	*D. chrysanthum*	10.	*E. nuda*
4.	*D. fibriatum*	11.	*E. ramentacea*
5.	*D. nobile*	12.	*Gnetum montanum*
6.	*Eulophia dabia*	13.	*Gymnadenia orchidis*
7.	*N. khasiana*	14.	*Nepenthes distillatoria*

Spices

India has a long history of producing and exporting spices. The world trade in spices is estimated around 0.45 million tonnes. In India, spices exports have been consistently moving up during the last one-decade with an increase of 210% in quantity and 622% in value during this period. The country commands 46% in global trade in terms of quantity and 28% in terms of value. Exports during the year 1998-99 have been 231389 tonnes valued at Rs.17580.2 million (Table 1.10). The exports rose during 1999-2000 creating an all time record in terms of value both in rupee and dollars. Compared to last year, export has registered an increase of 6% in rupee and dollar terms. However the export has shown a decrease of 10% in volume. The distribution of spice exports during 1991-92 and 1998-99 are given in Fig. 3.2 and Fig.3.3 respectively. During 1998-99, in the total spices export earnings, pepper contributed about 36.3 per cent followed by spice oils and oleoresins (24.1%) and chillies (12.3%) in terms of value. The value-added products in the export basket constitute 37% of the total value of exports.

Table 1.10 Estimated export of spices from India

Item	1991-92			1998-99		
	Quantity (tonnes)	Value (Rs. in millions)	% of total value	Quantity (tonnes)	Value (Rs. in millions)	% of total value
Pepper	20535	743.17	19.51	34864	6381.13	36.30
Cardamom (S)	544	155.74	4.09	475	252.12	1.43
Cardamom (L)	910	50.45	1.32	1424	119.09	0.68
Chillies	32603	894.85	23.49	61253	2166.11	12.32
Ginger	14259	218.81	5.74	8778	406.48	2.31
Turmeric	19661	377.62	9.91	36522	1245.50	7.08
Coriander	9954	132.35	3.47	20685	458.90	2.61
Cumin	1654	63.75	1.67	10723	601.09	3.42
Celery	3489	58.46	1.53	3991	96.91	0.55
Fennel	2136	48.09	1.26	5279	153.81	0.87
Fenugreek	6375	55.73	1.46	10082	191.49	1.09
Other seeds(1)	1282	25.15	0.66	2001	74.91	0.43
Garlic	10282	82.84	2.17	4068	74.10	0.42
Other spices (2)	13512	183.11	4.81	19077	765.95	4.36
Curry powder	3516	110.01	2.89	5210	359.68	2.05
Mint oil			0.00	4207	1225.22	6.97

Item	1991-92			1998-99		
	Quantity (tonnes)	Value (Rs. in millions)	% of total value	Quantity (tonnes)	Value (Rs. in millions)	% of total value
Spice oleoresins & other oils	1392	609.55	16.00	2750	3007.74	17.11
Total	142104	3810		231389	17580.23	

(1) Include aniseed, ajwain seed, dill seed, poppy seed, mustard etc.

(2) Include tamarind, asafoetida, cinnamon, cassia, kokam, saffron etc. Source: Spices Board, Kochi

PLANTATION CROPS

Coconut and its Products

While coconut as such is not exported, of late, the export of coconut products has increased considerably. The overall export of coconut and its products including coir increased from Rs. 731.90 million during 1991-92 to Rs. 3121.53 million in 1998-99.

Coconut oil fetched maximum export earnings of Rs. 62.19 million among the products other than coir, followed by coconut shell (raw)(Rs. 11.19 million), desiccated coconut (Rs. 8.92 million) and shell charcoal (Rs. 6.42 million). Coconut oil is exported primarily to Middle East and African countries like Jordan, Kuwait, Oman, UAE and Kenya.

Table 1.11 Export and import of coconut products and coir products

(Rs. in million)

Year	Exports of Coconut products	Coir Products	Import of Coconut Products
1991-92	4.30	627.60	21.92
1998-99	99.83	3021.70	43.28

There is still vast scope to increase export earnings from coconut and its products. Coir and coir products alone earned Rs. 2921 million during 1998-99.

Among the coconut products imported into India only refined coconut oil is imported from Indonesia and Philippines. The value of imports during 1998-99 was Rs. 43.28 million.

Arecanut

While it is claimed that arecanut production in the country is sufficient for internal consumption, a review of the available data provides some important trends. India has been a regular importer of arecanut and its exports, which was 658 tonnes valued at Rs. 45.70 million in 1991-92 has declined marginally during 1998-99 when only 533 tonnes of arecanut valued at Rs. 46.89 million was exported. During 1998-99, Indian imports were more than 12 times of exports, being 6707 tonnes valued at Rs. 187.56 million.

Table 1.12 Export and Import of Arecanut from India

Sl. No.	Year	Export		Import	
		Qty (tonnes)	Value (million Rs.)	Qty (tonnes)	Value (million Rs.)
1	1991-92	658	45.70	-	-
2	1998-99	533	46.89	6707	187.56

To reduce imports we have to lay emphasis in increasing productivity in the existing arecanut plantation rather than increasing the area under the crop. The arecanut imports were primarily from Sri Lanka followed by Indonesia and Myanmar. Small quantities of arecanut were imported once in a while from Hongkong, Bangladesh, Singapore, Bhutan, Nepal and Pakistan.

Oil Palm

The Country is importing huge quantity of vegetable oils, of which palm oil accounted for nearly 1.2 million tonnes. There is huge demand for vegetable oil in general and palm oil in particular, being a cheaper oil. The country has just started producing a small quantity of palm oil. There is, however, great scope for producing palm oil in the country, which can be exploited bringing one million ha under oil palm to produce three million tonnes of crude palm oil and 0.3 million tonnes of kernel oil by 2025.

Cashew

India is the leading producer, processor and exporter of cashew kernels in the world. The export earnings from cashew and allied products during 1991-92 was only Rs. 6690.90 million which increased to Rs. 16300.8 million by 1998-99. During 1999- 2000 the export rose to Rs. 24514 million, which is an all time record. Cashew stood fourth in position amongst horticultural products exported from India, after tea, coffee and spices. The export of cashew and its products has increased from 47,738 tonnes in 1991-92 to 77,026 tonnes in 1998-99. During the same period the value of exports increased from Rs. 6,690.90 million to Rs.16300.8 million. There is still potential for more exports in future.

Cashew kernels obtained from raw cashewnuts are exported to more than 30 countries. Very small quantities of roasted/salted cashewnuts in consumer packs of 1 kg or less are also exported. Principal destinations of exports are the USA, the Netherlands, Japan, U.K, Australia, Singapore, France, UAE and Germany.

Cashew nut shell liquid (CNSL) is also an exported item, which was 4542 Mt. in 1991-92 dripped to 1912 Mt in 1998-99, which earned RS, 42.1 million. The details of export and import of cashew is given in Table 1.13.

The growth of cashew processing industry has been substantial over the last two decades, whereby the processing units doubled, doubling the domestic production. There were 572 units in early 80's, consuming nearly 0.35 million tonnes of raw nuts. This increased to 1098 units in 1998-1999, consuming nearly 0.65 million tonnes of raw nuts. The processing capacity of these units is about 1 million tonnes, which is not being fully utilized due to short supply of raw nuts. The average demand growth rate for cashew kernels for export and internal consumption is around 13 per cent. However, the production of raw cashewnuts in the country at about 0.35 million tonnes is inadequate to meet the requirements of the installed processing capacity in the country. The cashew processors are, therefore, importing raw cashewnuts to meet their demand. During 1998-99, about 0.24 million tonnes of raw nuts worth Rs. 9580.3 million were imported.

Keeping in view the export potential of cashew, several countries are laying emphasis on cashew cultivation. Particularly countries like Brazil, Tanzania, Vietnam and Indonesia are emerging as competitors. Brazil and Tanzania are already close to 1/4th of the total cashew production in India. Some of these countries are improving their processing capability. To maintain its leadership in exporting cashew, India will have to increase its production to meet the installed processing capacity within the domestic market, improve the quality and reduce cost of production to be competitive.

Table 1.13 Export of cashew kernel, cashewnut shell liquid (CNSL) and import of raw cashewnut during 1990-91 to 1998-99

Value in Rs. in Million Quantity in MT

	Cashew Kernel		CNSL		Raw nut	
Year	Quantity	Value	Quantity	Value	Quantity	Value
1991-92	47738	6690.9	4542	40.2	106080	2666.8
1992-93	53436	7454.9	4258	38.1	134985	3763.3
1993-94	69884	10460.2	3625	29.0	191322	4827.0
1994-95	77000	12460.2	3807	24.4	228109	6909.4
1995-96	70334	12405.0	760	14.5	222819	7600.8
1996-97	68663	12855.0	1735	27.7	212866	6876.0
1997-98	76593	13961.0	4446	71.7	247181	7696.0
1998-99	77076	16300.8	1912	42.1	241161	9580.3

Cocoa

Large quantity of cocoa are being imported as the present production of about 7000 tonnes is far less than the processing capacity of 30000 tonnes per annum. Efforts will have to be made to bridge the gap to reduce the foreign exchange loss.

Tea

Until 1987-88 tea was the most significant export item from India. It accounted for 20.7% of total export and ranked first among the agricultural products. Thereafter, the export has declined considerably third position by contributing only 8.8% of the total exports.

The details of export of tea are given in Table 1.14. In 1990 the share of exports decreased to 18.55, whole in 1999 it further came down to 14.31. Whole developing markets stagnate growth is expected in the developing countries. These markets are expected to open up substantially in the wake of the WTO regime. This will however bring about increased global competitiveness in the area of cost, price, quality, supply schedules, packaging and market focus and customer satisfaction.

Table 1.4 Export of Tea from India

Year	Export (Qty m Kg)	Value Rs. million	Unit Price Rs. /Kg
1950	200.7	804.2	4.10
1991	202.9	11345.5	55.91
1999	191.7	19658.6	102.5

The overall tea situation in India is quite different from other producing countries. While some of these countries like Sri Lanka export as much as 95 per cent of their produce, in case of India nearly 75 percent of the total production is consumed within the country. However, the per capita consumption of tea in India is still one of the lowest in the world with only 660 gms per head. With improved economic levels, the demand for domestic consumption is expected to increase. As a consequence India has been importing tea from other countries. During 1999 India imported 9.79 million kgs of tea valued at Rs.590.70 million.

The issues affecting India's export performance in tea are:-

- Inadequate efforts made in respect of marketing and brand building in overseas market.
- Over dependence for export on CIS countries, UK and UAE.
- Competitions from countries like Sri Lanka and in Russia.
- Limited presence in emerging market like Syria, Germany, Kenya, Saudi Arabia and Dubai.
- Higher growth rate registered by China in green tea marketing.

The export of tea from India during 1999 was estimated at 191.7 million kg.valued at Rs. 19658.6 million with a unit price of Rs.102.5 per kg. The increase in value of exports between 1991 and 1999 indicates a decline in export volume. On the other hand the export earnings have gone up in value terms. The Indian share in global tea exports, however had declined from considerably during the period.

Import of tea touched around 9.79 million in 1999 compared to 1-2 million kg upto 1997-98 (Table 1.15). This was value at Rs. 590.7 million at a unit CIF value of Rs. 60.39. The unit CIF value of tea imported from a few countries was lower than the prices fetched at Indian auction. As a result Indian tea industry which was earlier competing in the international market will now have to compete in domestic market also protecting the home market is going to be a challenging task. The Indian tea industry will have to evolve same strategy to counter the new competing forces unleashed by globalization.

Table: 1.15 Import of tea into India during 1995-2000

Year	Quantity (million kg)	Value (million Rs.)	Unit (Rs./kg)
1995-96	0.45	24.1	53.56
1996-97	1.25	62.1	49.68
1997-98	2.61	177.9	68.16
1998-99	8.93	647.3	72.79

Coffee

India's coffee industry has great potential for exports and 80 per cent of total produce is exported to over 60 countries worldwide. The major importing countries are Italy, Germany and Russian Federation which together accounts for 43.93 per cent of imports from India.

The countries like USA, Germany, Spain and Belgium are the major buyers of arabica coffee, while Russian Federation imports mainly the other value added products (instant coffee). The emerging new markets for Indian coffee are Japan and Middle East countries. India's share in world exports is around 4.6 per cent. India exports both arabica, which is classified under "other mild category" and robusta with their percentage share in world exports of 5 and 12 per cent respectively. During 1998-99, exports amounted to 2,11,623 tonnes earning Rs.17515 million (US $ 431.4 million) (Table 1.16).

Table 1.16 Export a coffee from India

Year	Export Q/Mt	Value (RS. in million)	Unit price (RS. /Kg)
1990-91	100110	2788.90	27.58
1998-99	211623	17515.30	82.77

India's share in world export of coffee was only 2.0 and 4.7 per cent during 1991-92 and 1998-99 respectively. During the X Plan an annual coffee growth rate of 5 per cent is envisaged keeping a base level of 2,50,000 tonnes.

India has to evolve suitable strategy to increase exports in the current oversupply situation. Even through India produces the best robusta in the world, it will find it difficult to compete with its Asian neighbours, i.e. Vietnam and Indonesia in the cost of production. However, Indian Arabica is more cost competitive than others, and India should exploit this situation through engaging the key market of USA for Indian coffee. Similarly in Germany the focus should be in building washed Arabica.

Rubber

The accelerated growth in demand for natural rubber in India made the country a net importer of Natural rubber in 1947. The situation remained unchanged until 1970. But, during the first three years of 1970s, owing to the industrial recession and slackness in demand, import of natural rubber was banned from April 1973. Apart from this measure, limited quantities of natural rubber were exported during 1973-74, 1974- 75,1976-77 and 1977-78. To overcome occasional glut in the market and also in accordance with the new economic policy of Government of India, all restriction on export of natural rubber was removed since 1991. But the indigenous production was insufficient to meet internal consumption and natural rubber continued to be deficit in the country. Though export was freely allowed, only nominal quantities were actually exported. However, India has been exporting significant quantities of rubber in the form of value added products. The export earning from rubber products during 1999-2000 was Rs. 5730 million.

Factor – Factor Relationship

While cultivating crops or managing allied enterprises, a farmer has several options to combine some inputs, i.e., to meet the nutrient requirements for crops, the farmer can use a combination of organic and inorganic fertilizers, while controlling pests he can use bio-control and chemical control, combination of animal and machine power, combination of green fodder and concentrates in the case of livestock etc. The farmer has to use the optimum or least cost combination of two or more inputs for producing a given amount of output. The combining of inputs will be possible only if the inputs considered are substitutable. Therefore it is essential to examine the possibilities of substituting one input (X1) for another input (X2) while the output level (Y) is held constant.

Y=f X1, X2 / X3, X4, ….. Xn). Here Y is a function of X1 and X2 while other inputs are kept constant. This expression shows that the amount of output (Y) depends in a unique way upon the amount of the two inputs X1 and X2 used in the production process along with the other fixed inputs.

Isoquant or Iso-product Curve

Isoquant is the line connecting all combinations of two inputs that would produce the same level of output. Each point on an isoquants is the maximum output that can be achieved with these input combinations.

The data on various combinations of nitrogenous and phosphorous nutrients applied to brinjal and the corresponding yield is given in Table 1.17.

Table 1.17 Estimated Yields (Qtls) for Various Combinations of Nitrogen and Phosphorus for Brinjal

P/N	40	60	80	100	120	140
40	40.21	43.09	44.74	45.15	44.32	44.02
60	41.87	45.51	47.90	49.06	48.98	48.32
80	40.89	45.28	48.43	50.35	51.02	50.54

Types of Factor-Factor Relationships

There can be three types of combinations of inputs:

(i) Fixed proportion combinations of inputs

(ii) Constant rate of substitution

(iii) Varying rates of substitution

(i) Fixed Proportion Combinations

Some of the inputs can be combined only in fixed proportions. Hence, in such cases, there is no decision-problem.

(ii) Constant Rate of Substitution

The substitution at constant rate occurs when each unit of input (X1) added every time would replace the same level of other input (X2).

Substitution Ratio = Δ X21 = Δ X22 = Δ X2n

Δ X11 = Δ X12 = Δ X1n

(iii) Varying Rate of Substitution

The amount of one input (X1) required to substitute for one unit of another input (X2) for a given level of production increases or decreases as the amount of X2 used increases. Substitution at an increasing rate is not common in horticulture, but decreasing rate of substitution is more common. The slope of the iso-product curve in this case becomes less steep as more of X1 is used relative to X2.

Substitution Ratio change as under:

Δ X21 Δ X22 Δ X2n

\> >.....>

Δ X11 Δ X12 Δ X1n

Iso – cost Lines

An iso-cost line is the line connecting all combinations of two inputs that could be purchased for a given budget.

Least- cost combination

Simple Arithmetical Calculations

One possible way to determine the least-cost combination is to compute the cost of all possible combinations and then select the one with the minimum cost. This method is suitable where only a few combinations produce a given output and calculations involved are a few and simple. Suppose there are five combinations of inputs which can produce 85 units of output as given in table 1.18. The price per unit of X1 is Rs.3 and of X2 Rs.4. The total outlay of each combination can be determined by simple multiplications.

Table 1.18 Least-cost Combination of Two Inputs for Producing an Output

Units of X1	Units of X2	Cost of X1 @ Rs.3.00	Cost of X2 @ Rs.4/-	Total outlay for X1 and X2 (Rs.)
8.0	2	24.00	8	32.0
6.0	3	18.00	12	30.0
5.0	4	15.00	16	31.0
4.5	5	13.50	20	33.5
3.5	7	10.50	28	38.5

Out of five combinations calculated in table 1.18 above, 3 units of X2 and 6 units of X1 is the least cost combination of inputs i.e., Rs.30. through this method it is possible that the true least cost combination is not located because we do not consider many combination which may be there on the iso-quant to the right or to the leaf of the point.

Graphic Method

Isoquants and iso-cost lines are drawn on the same graph for different levels of production and total outlay. The least cost combination is indicated by the point of tangency of the isoquants and iso-cost lines.

Product – product relationship

The farmer has limited resources with which he has to maximize the revenue from the farm. Therefore, it is essential for the farmer to decide the crops or enterprises that he has to undertake so as to maximize the farm income. The nature of combination of enterprises and the rate of product substitution will influence the optimum combination of enterprises.

Production possibility curve (PPC) or Iso-resource Curve

The Production Possibility Curve is a line connecting all combinations outputs of two enterprises or crops that can be produced using a given amount of input.

Marginal Rate of product substitution (MRPS)

MRPS is the rate of change in quantity of output of one enterprise (Y1) as a result of unit increase in the output of the other enterprise (Y2), for a given level of input (X). It is also known as Marginal Rate of Product Transformation (MRPT). So it is the slope of the PPC.

MRPS = DY1/DY2

Types of Product - Product relationship

Competitive products

Two products (or enterprises) are competitive when the output of one product can be increased only by reducing the output of the other product. Outputs are competitive because they require the same inputs at the same time. If a farmer has a given level of water for irrigating bhendi and brinjal, he can either allocate equal share of water to both or more of water to bhendi and less to brinjal. Hence, these two products become competitive.

Complementary product

Two products (or enterprises) are complementary if an increase in output of one product (Y1) also causes an increase in the output other product (Y2) also, for the same level of inputs. PPC for complementary products have positive slope. Eg: Rotation of complimentary crop in cropping pattern. The by-products of one complementary enterprise (Y1) will serve as input for production of the other product (Y2).

Supplementary products

Two products are called supplementary if one product can be increased without increasing or decreasing the other product. These two enterprises are not interlinked. Eg: Farm yard Poultry and Crop cultivation.

Joint products

Products which result from the same production process are termed as joint products. No substitution among products is possible since joint products are produced in a fixed proportion. Eg: cotton lint and cotton seed.

MRPS and enterprise relation ships

MRPS	Enterprise relationship
DY2/DY1 > O	Complementary
DY2/DY1 > O	Competitive
DY2/DY1 > O	Supplementary

Iso-Revenue line or price line

It is a line which connects all the possible combinations of two commodities which would yield a given level of revenue or income.

2

AGRICULTURAL MARKETING

Agricultural marketing is inferred to cover the services involved in moving an agricultural product from the farm to the consumer. It is also the planning, organizing, directing and handling of agricultural produce in such a way as to satisfy the farmer, producer and the consumer. Numerous interconnected activities are involved in doing this, such as planning production, growing and harvesting, grading, packing and packaging, transport, storage, agro- and food processing, distribution, advertising and sale. Effectively, the term encompasses the entire range of supply chain operations. However, its key function is to help direct these services, by providing competent and able market information, thereby linking the other operations into an integrated service with targeted outcomes.

AGRICULTURAL MARKETING DEVELOPMENT

Efforts to develop agricultural marketing have, particularly in developing countries, tended to concentrate on a number of areas, specifically infrastructure development; information provision; training of farmers and traders in marketing and post-harvest issues; and support to the development of an appropriate policy environment. In the past, efforts were made to develop government-run marketing bodies but these have tended to become less prominent over the years.

Agricultural Market infrastructure

Efficient marketing infrastructure such as wholesale, retail and assembly markets and storage facilities is essential for cost-effective marketing, to minimize post-harvest losses and to reduce health risks. Markets play an important role in rural development, income generation, food security, and developing rural-market linkages. Experience shows that planners need to be aware of how to design markets that meet a community's social and economic needs and how to choose a suitable site for a new market. In many cases sites are chosen that are inappropriate and result in under-use or even no use of the infrastructure constructed. It is also not sufficient just to build a market: attention needs to be paid to how that market will be managed, operated and maintained.

Rural assembly markets are located in production areas and primarily serve as places where farmers can meet with traders to sell their products. These may be occasional (perhaps weekly) markets, such as haat bazaars in India and Nepal, or permanent. Terminal wholesale markets are located in major metropolitan areas, where produce is finally channelled to consumers through trade between wholesalers and retailers, caterers, etc. The characteristics of wholesale markets have changed considerably as retailing changes in response to urban growth, the increasing role of supermarkets and increased consumer spending capacity. These changes may require responses in the way in which traditional wholesale markets are organized and managed.

Retail marketing systems in western countries have broadly evolved from traditional street markets through to the modern hypermarket or out-of-town shopping center. In developing countries, there remains scope to improve agricultural marketing by constructing new retail markets, despite the growth of supermarkets, although municipalities often view markets primarily as sources of revenue rather than infrastructure requiring development. Effective regulation of markets is essential. Inside a market, both hygiene rules and revenue collection activities have to be enforced. Of equal importance, however, is the maintenance of order outside the market. Licensed traders in a market will not be willing to cooperate in raising standards if they face competition from unlicensed operators outside who do not pay any of the costs involved in providing a proper service.

Market information

Efficient market information can be shown to have positive benefits for farmers and traders. Up-to-date information on prices and other market factors enables farmers to negotiate with traders and also facilitates spatial distribution of products from rural areas to towns and between markets. Most governments in developing countries have tried to provide market information services to farmers, but these have tended to experience problems of sustainability. Moreover, even when they function, the service provided is often insufficient to allow commercial decisions to be made because of time lags between data collection and dissemination. Modern communications technologies open up the possibility for market information services to improve information delivery through SMS on cell phones and the rapid growth of FM radio stations in many developing countries offers the possibility of more localised information services.

In the longer run, the internet may become an effective way of delivering information to farmers. However, problems associated with the cost and accuracy of data collection still remain to be addressed. Even when they have access to market information, farmers often require assistance in interpreting that information. For example, the market price quoted on the radio may refer to a wholesale selling price and farmers may have difficulty in translating this into a realistic price at their local assembly market. Various attempts have been made in developing countries to introduce commercial market information services but these have largely been targeted at traders, commercial farmers or exporters. It is not easy to see how small, poor farmers can generate sufficient income for a commercial service to be profitable although in India a new service introduced by Thomson Reuters was reportedly used by over 100,000 farmers in its first year of operation. Esoko in West Africa attempts to subsidize the cost of such services to farmers by charging access to a more advanced feature set of mobile-based tools to businesses.

Marketing training

Farmers frequently consider marketing as being their major problem. However, while they are able to identify such problems as poor prices, lack of transport and high post-harvest losses, they are often poorly equipped to identify potential solutions. Successful marketing requires learning new skills, new techniques and new ways of obtaining information. Extension officers working with ministries of agriculture or NGOs are often well-trained in agricultural production techniques but usually lack knowledge of marketing or post-harvest handling.

Enabling environments

Agricultural marketing needs to be conducted within a supportive policy, legal, institutional, macro-economic, infrastructural and bureaucratic environment. Traders and others are generally reluctant to make investments in an uncertain policy climate, such as those that restrict imports and exports or internal produce movement. Businesses have difficulty functioning when their trading activities are hampered by excessive bureaucracy. Inappropriate law can distort and reduce the efficiency of the market, increase the costs of doing business and retard the development of a competitive private sector. Poor support institutions, such as agricultural extension services, municipalities that operate markets inefficiently and inadequate export promotion bodies, can be particularly damaging. Poor roads increase the cost of doing business, reduce payments to farmers and increase prices to consumers. Finally, corruption can increase the transaction costs faced by those in the marketing chain.

Agricultural marketing support

Most governments have at some stage made efforts to promote agricultural marketing improvements. In the United States the Agricultural Marketing Service (AMS) is a division of USDA and has programs that provide testing, support standardization and grading and offer market news services. AMS oversees marketing agreements and orders research and promotion programs. It also purchases commodities for federal food programs. USDA also provides support to agricultural marketing work at various universities. In the United Kingdom, support for marketing of some commodities was provided before and after the Second World War by boards such as the Milk Marketing Board and the Egg Marketing Board. These boards were closed down in the 1970s. As a colonial power, Britain established marketing boards in many countries, particularly in Africa. Some continue to exist although many were closed at the time of the introduction of structural adjustment measures in the 1990s.

Several developing countries have established government-sponsored marketing or agribusiness units. South Africa, for example, started the National Agricultural Marketing Council (NAMC) as a response to the deregulation of the agriculture industry and closure of marketing boards in the country. India has the long-established National Institute of Agricultural Marketing. These are primarily research and policy organizations, but other agencies provide facilitating services for marketing channels, such as the provision of infrastructure, market information and documentation support. Examples from the Caribbean include the National Agricultural Marketing Development Corporation (NAMDEVCO) in Trinidad and Tobago and the New Guyana Marketing Corporation in Guyana.

Recent developments

New marketing linkages between agribusiness, large retailers and farmers are gradually being developed, e.g. through contract farming, group marketing and other forms of collective action. Donors and NGOs are paying increasing attention to ways of promoting direct linkages between farmers and buyers within a value chain context. More attention is now being paid to the development of regional markets (e.g. East Africa) and to structured trading systems that should facilitate such developments. The growth of supermarkets, particularly in Latin America and East and South East Asia, is having a significant impact on marketing channels for horticultural, dairy and livestock products. Nevertheless, "spot" markets will continue to be important for many years, necessitating attention to infrastructure improvement such as for retail and wholesale markets.

Energy crop

An energy crop is a plant grown as a low-cost and low-maintenance harvest used to make biofuels, such as bioethanol, or combusted for its energy content to generate electricity or heat. Energy crops are generally categorized as woody or herbaceous plants; many of the latter are grasses of the family Graminaceae.

Commercial energy crops are typically densely planted, high-yielding crop species which are processed to bio-fuel and burnt to generate power. Woody crops such as willow or poplar are widely utilised, as well as temperate grasses such as Miscanthus and Pennisetum purpureum (both known as elephant grass). If carbohydrate content is desired for the production of biogas, whole-crops such as maize, Sudan grass, millet, white sweet clover, and many others can be made into silage and then converted into biogas.

Through genetic modification and application of biotechnology plants can be manipulated to create greater yields, high energy yields can also be realized with existing cultivars.. However, some additional advantages such as reduced associated costs (i.e. costs during the manufacturing process) and less water use can only be accomplished by using Genetically_modified_crops#Biofuel genetically modified crops.

Types

Energy generated by burning plants grown for the purpose, often after the dry matter is pelletized. Energy crops are used for firing power plants, either alone or co-fired with other fuels. Alternatively they may be used for heat or combined heat and power (CHP) production.

To cover the increasing requirements of woody biomass, short rotation coppice (SRC) were applied to agricultural sites. Within this cropping systems fast growing tree species like willows and poplars are planted in growing cycles of three to five years. The cultivation of this cultures is dependent on wet soil conditions and could be an alternative for moist field sieds. However, an influence on local water conditions could not be excluded. This indicates that an establishment should exclude the vicinity to vulnerable wetland ecosystems.

Gas biomass (methane)

Anaerobic digesters or biogas plants can be directly supplemented with energy crops once they have been ensiled into silage. The fastest growing sector of German biofarming has been in the area of "Renewable Energy Crops" on nearly 500,000 ha (1,200,000 acres) of land (2006). Energy crops can also be grown to boost gas yields where feedstocks have a low energy content, such as manures and spoiled grain. It is estimated that the energy yield presently of bioenergy crops converted via silage to methane is about 2 GWh/km (1.8×10 BTU/sq mi). Small mixed cropping enterprises with animals can use a portion of their acreage to grow and convert energy crops and sustain the entire farms energy requirements with about one fifth of the acreage. In Europe and especially Germany, however, this rapid growth has occurred only with substantial government support, as in the German bonus system for renewable energy. Similar developments of integrating crop farming and bioenergy production via silage-methane have been almost entirely overlooked in N. America, where political and structural issues and a huge continued push to centralize energy production has overshadowed positive developments.

Liquid biomass

European production of biodiesel from energy crops has grown steadily in the last decade, principally focused on rapeseed used for oil and energy. Production of oil/biodiesel from rape covers more than 12,000 km^2 in Germany alone, and has doubled in the past 15 years.Typical yield of oil as pure biodiesel may be is 100,000 L/km (68,000 US gal/sq mi; 57,000 imp gal/sq mi) or more, making biodiesel crops economically attractive, provided sustainable crop rotations exist that are nutrient-balanced and preventative of the spread of disease such as clubroot. Biodiesel yield of soybeans is significantly lower than that of rape.

Bioethanol

Energy crops for biobutanol are grasses. Two leading non-food crops for the production of cellulosic bioethanol are switchgrass and giant miscanthus. There has been a preoccupation with cellulosic bioethanol in America as the agricultural structure supporting biomethane is absent in many regions, with no credits or bonus system in place. Consequently, a lot of private money and investor hopes are being pinned on marketable and patentable innovations in enzyme hydrolysis and the like.

Bioethanol also refers to the technology of using principally corn (maize seed) to make ethanol directly through fermentation, a process that under certain field and process conditions can consume as much energy as is the energy value of the ethanol it produces, therefore being non-sustainable. New developments in converting grain stillage (referred to as distillers grain stillage or DGS) into biogas energy looks promising as a means to improve the poor energy ratio of this type of bioethanol process.

By dedication

Dedicated energy crops are non-food energy crops as giant miscanthus, switchgrass, jatropha, fungi, and algae. Dedicated energy crops are promising cellulose sources that can be sustainably produced in many regions of the United States. Additionally, the green waste byproducts of food and non-food energy crops can be used to produce various biofuels.

Agricultural value chain

The agricultural value chain concept has been used since the beginning of the millennium, primarily by those working in agricultural development in developing countries. Although there is no universally accepted definition of the term, it normally refers to the whole range of goods and services necessary for an agricultural product to move from the farm to the final customer or consumer.

The term value chain was first popularized in a book published in 1985 by Michael Porter, who used it to illustrate how companies could achieve what he called "competitive advantage" by adding value within their organization. Subsequently, the term was adopted for agricultural development purposes and has now become very much in vogue among those working in this field, with an increasing number of bilateral and multilateral aid organisations using it to guide their development interventions.

At the heart of the agricultural value chain concept is the idea of actors connected along a chain producing and delivering goods to consumers through a sequence of activities. However, this "vertical" chain cannot function in isolation and an important aspect of the value chain approach is that it also considers "horizontal" impacts on the chain, such as input and finance provision, extension support and the general enabling environment. The approach has been found useful, particularly by donors, in that it has resulted in a consideration of all those factors impacting on the ability of farmers to access markets profitably, leading to a broader range of chain interventions. It is used both for upgrading existing chains and for donors to identify market opportunities for small farmers.

Definitions

There is no commonly agreed definition of what is actually meant by agricultural value chains. Indeed, some agencies are using the term without having a workable definition or definitions and simply redefined ongoing activities as "value chain" work when the term came into vogue. Published definitions include the World Bank's "the term "value chain" describes the full range of value adding activities required to bring a product or service through the different phases of production, including procurement of raw materials and other inputs", UNIDO's "actors connected along a chain producing, transforming and bringing goods and services to end-consumers through a sequenced set of activities", and CIAT's "a strategic network among a number of business organizations".

Without a universal definition the term "value chain" is now being used to refer to a range of types of chain, including:

- An international, or regional commodity market. Examples could include "the global cotton value chain", "the southern African maize value chain" or "the Brazilian coffee value chain";
- A national or local commodity market or marketing system such as "the Ghanaian tomato value chain" or ""the Accra tomato value chain";
- A supply chain, which can cover both of the above;
- An extended supply chain or marketing channel, which embraces all activities needed to produce the product, including information/ extension, planning, input supply and finance. It is probably the most common usage of the value chain term;
- A dedicated chain designed to meet the needs of one or a limited number of buyers. This usage, which is arguably most faithful to Porter's concept, stresses that a value chain is designed to capture value for all actors by carrying out activities to meet the demand of consumers or of a particular retailer, processor or food service company supplying those consumers. Emphasis is firmly placed on demand as the source of the value.

VALUE CHAIN METHODOLOGIES

Donors and others supporting agricultural development, such as GIZ, DFID, ILO, IIED and UNIDO, have produced a range of documents designed to assist their staff and others to evaluate value chains in order to decide on the most appropriate interventions to either update existing chains or promote new ones. However, the application of value chain analysis is being interpreted differently by different organisations, with possible repercussions for their development impact. The proliferation of guides has taken place in an environment where key conceptual and methodological elements of value chain analysis and development are still evolving. Many of these guides include not only detailed procedures that require experts to carry out the analysis but also use detailed quasi-academic methodologies. One such methodology is to compare the same value chain over time (a comparative or panel study) to assess changes in rents, governance, systemic efficiency and the institutional framework.

Linking farmers to markets

A major subset of value chain development work is concerned with ways of linking producers to companies, and hence into the value chains. While there are examples of fully integrated value chains that do not involve smallholders (e.g. Unilever operates tea estates and tea processing facilities in Kenya and then blends and packs the tea in Europe before selling it as Lipton, Brooke Bond or PG Tips brands), the great bulk of agricultural value chains involve sales to companies from independent farmers. Such arrangements frequently involve contract farming in which the farmer undertakes to supply agreed quantities of a crop or livestock product, based on the quality standards and delivery requirements of the purchaser, often at a price that is established in advance. Companies often also agree to support the farmer through input supply, land preparation, extension advice and transporting produce to their premises.

Inclusive value chains

Work to promote market linkages in developing countries is often based on the concept of "inclusive value chains", which usually places emphasis on identifying possible ways in which small-scale farmers can be incorporated into existing or new value chains or can extract greater value from the chain, either by increasing efficiency or by also carrying out activities further along the chain. In the various publications on the topic the definition of "inclusion" is often imprecise as it is often unclear whether the development aim is to include all farmers or only those best able to take advantage of the opportunities.

Agricultural value chain finance

Agricultural value chain finance is concerned with the flows of funds to and within a value chain to meet the needs of chain actors for finance, to secure sales, to buy inputs or produce, or to improve efficiency. Examining the potential for value chain finance involves a holistic approach to analyze the chain, those working in it, and their inter-linkages. These linkages allow financing to flow through the chain. For example, inputs can be provided to farmers and the cost can be repaid directly when the product is delivered, without need for farmers taking a loan from a bank or similar institution. This is common under contract farming arrangements. Types of value chain finance include product financing through trader and input supplier credit or credit supplied by a marketing company or a lead firm. Other trade finance instruments include receivables financing where the bank advances funds against an assignment of future receivables from the buyer, and factoring in which a business sells its accounts receivable at a discount. Also falling under value chain finance are asset collateralization, such as on the basis of warehouse receipts, and risk mitigation, such as forward contracting, futures and insurance.

The use of ICTs in value chains

Information and Communication Technologies, or ICTs, have become an important tool in promoting agricultural value chain efficiency. There has been a rapid expansion in the use of mobile technologies, in particular. The price of ICT services is falling and the technologies are becoming more affordable to many in developing countries. Applications can support farmers directly through SMS messages. Examples include iCow, developed in Kenya, which provides information on the gestation period, on artificial insemination of the cows, and on how to look after them. Applications such as M-Pesa can support access to mobile payment services for a large percentage of those without banks, thereby facilitating transactions in the value chain. Other applications have been developed to promote provision of crop insurance through input dealers, for example.

ICTs are also being used to strengthen the capacity of extension officers and NGO field staff to reach farmers with timely and accurate information and, at the same time, help capture data from the field. The Grameen Foundation's Community Knowledge Worker (CKW) programme is an example. Farmer representatives are trained to use ICT applications on a smartphone to provide agricultural information and extension support. Most market price information is now delivered to farmers via SMS. Further along the chain, technologies offer considerable possibilities to enhance traceability, which is particularly relevant as certification grows in importance. Where necessary many exporters can now trace consignments back to individual farmers and take necessary measures to address problems. Finally, systems such as FARA's eRails are also supporting agricultural researchers through data collection and analysis and access to up-to-date research publications.

Enabling environments

As with all agricultural growth, two things appear essential for successful value chain development: creating the right environment for agriculture and investing in rural public goods. An enabling environment implies peace and public order, macro-economic stability, inflation under control, exchange rates based on market fundamentals rather than government allocation of foreign currency, predictable taxation that is reinvested in public goods and property rights. There is a positive correlation of agricultural growth with investment in irrigation, transport infrastructure and other technologies. Governments have a responsibility to provide essential goods and services, infrastructure, such as rural roads, and agricultural research and extension. Value chain development is often constrained by corruption, both at a high level and at the ubiquitous road blocks found in many countries, particularly in Africa. Many measures to improve value chains require collaboration between a wide range of different ministries, and this can be difficult to achieve.

Intensive farming

Intensive farming involves various types of agriculture with higher levels of input and output per cubic unit of agricultural land area. It is characterized by a low fallow ratio, higher use of inputs such as capital and labour, and higher crop yields per cubic unit land area.This contrasts with traditional agriculture, in which the inputs per unit land are lower. The term "intensive" involves various meanings, some of which refer to organic farming methods (such as biointensive agriculture and French intensive gardening), and others that refer to nonorganic and industrial methods. Intensive animal farming involves either large numbers of animals raised on limited land, usually concentrated animal feeding operations (CAFOs), often referred to as factory farms, or managed intensive rotational grazing(MIRG), which has both organic and non-organic types. Both increase the yields of food and fiber per acre as compared to traditional animal husbandry. In CAFO, feed is brought to the seldom-moved animals, while in MIRG the animals are repeatedly moved to fresh forage.

Most commercial agriculture is intensive in one or more ways. Forms that rely heavily on industrial methods are often called industrial agriculture, which is characterised by innovations designed to increase yield. Techniques include planting multiple crops per year, reducing the frequency of fallow years, and improving cultivars. It also involves increased use of fertilizers, plant growth regulators, and pesticides and mechanised agriculture, controlled by increased and more detailed analysis of growing conditions, including weather, soil, water, weeds, and pests. This system is supported by ongoing innovation in agricultural machinery and farming methods, genetic technology, techniques for achieving economies of scale, logistics, and data collection and analysis technology. Intensive farms are widespread in developed nations and increasingly prevalent worldwide. Most of the meat, dairy, eggs, fruits, and vegetables available in supermarkets are produced by such farms.

Smaller intensive farms usually include higher inputs of labor and more often use sustainable intensive methods. The farming practices commonly found on such farms are referred to as appropriate technology. These farms are less widespread in both developed countries and worldwide, but are growing more rapidly. Most of the food available in specialty markets such as farmers markets is produced by these small holder farms.

Agricultural development in Britain between the 16th century and the mid-19th century saw a massive increase in agricultural productivity and net output. This in turn supported unprecedented population growth, freeing up a significant percentage of the workforce, and thereby helped enable the Industrial Revolution. Historians cited enclosure, mechanization, four-field crop rotation, and selective breeding as the most important innovations.

Industrial agriculture arose along with the Industrial Revolution. By the early 19th century, agricultural techniques, implements, seed stocks, and cultivars had so improved that yield per land unit was many times that seen in the Middle Ages.

The industrialization phase involved a continuing process of mechanization. Horse-drawn machinery such as the McCormick reaperrevolutionized harvesting, while inventions such as the cotton gin reduced the cost of processing. During this same period, farmers began to use steam-powered threshers and tractors, although they were expensive and dangerous. In 1892, the first gasoline-poweredtractor was successfully developed, and in 1923, the International Harvester Farmall tractor became the first all-purpose tractor, marking an inflection point in the replacement of draft animals with machines. Mechanical harvesters (combines), planters, transplanters, and other equipment were then developed, further revolutionizing agriculture. These inventions increased yields and allowed individual farmers to manage increasingly large farms.

The identification of nitrogen, phosphorus, and potassium (NPK) as critical factors in plant growth led to the manufacture of synthetic fertilizers, further increasing crop yields. In 1909, the Haber-Bosch method to synthesize ammonium nitrate was first demonstrated. NPK fertilizers stimulated the first concerns about industrial agriculture, due to concerns that they came with serious side effects such as soil compaction, soil erosion, and declines in overall soil fertility, along with health concerns about toxic chemicals entering the food supply.

The identification of carbon as a critical factor in plant growth and soil health, particularly in the form of humus, led to so-called sustainable agriculture, as well as alternative forms of intensive agriculture that also surpassed traditional agriculture, without side effects or health issues. Farmers adopting this approach were initially referred to as humus farmers, later as organic farmers.

The discovery of vitamins and their role in nutrition, in the first two decades of the 20th century, led to vitamin supplements, which in the 1920s allowed some livestock to be raised indoors, reducing their exposure to adverse natural elements. Chemicals developed for use in World War II gave rise to synthetic pesticides.

Following World War II, synthetic fertilizer use increased rapidly, while sustainable intensive farming advanced much more slowly. Most of the resources in developed nations went to improving industrial intensive farming, and very little went to improving organic farming. Thus, particularly in the developed nations, industrial intensive farming grew to become the dominant form of agriculture.

The discovery of antibiotics and vaccines facilitated raising livestock in confined animal feeding operations by reducing diseases caused by crowding. Developments in logistics and refrigeration as well as processing technology made long-distance distribution feasible.

Between 1700 and 1980, "the total area of cultivated land worldwide increased 466%" and yields increased dramatically, particularly because of selectively-bred, high-yielding varieties, fertilizers, pesticides, irrigation, and machinery. Global agricultural production doubled between 1820 and 1920; between 1920 and 1950; between 1950 and 1965; and again between 1965 and 1975 to feed a global population that grew from one billion in 1800 to 6.5 billion in 2002. The number of people involved in farming in industrial countries dropped, from 24 percent of the American population to 1.5 percent in 2002. In 1940, each farmworker supplied 11 consumers, whereas in 2002, each worker supplied 90 consumers. The number of farms also decreased and their ownership became more concentrated. In the year 2000 in the U.S., four companies produced 81 percent of cows, 73 percent of sheep, 57 percent of pigs, and 50 percent of chickens, which was cited as an example of "vertical integration" by the president of the U.S. National Farmers Union.Between 1967 and 2002, the one million pig farms in America consolidated into 114,000, with 80 million pigs (out of 95 million) produced each year on factory farms, according to the U.S. National Pork Producers Council. According to the Worldwatch Institute, 74 percent of the world's poultry, 43 percent of beef, and 68 percent of eggs are produced this way.

Concerns over the sustainability of industrial agriculture, which has become associated with decreased soil quality, and over the environmental effects of fertilizers and pesticides, have not subsided. Alternatives such as integrated pest management (IPM) have had little impact because policies encourage the use of pesticides and IPM is knowledge-intensive. These concerns sustained the organic movement and caused a resurgence in sustainable intensive farming, as well as funding for the development of appropriate technology.

Techniques and technologies

Intensive livestock farming, also called "factory farming", is a term referring to the process of raising livestock in confinement at high stocking density. "Concentrated animal feeding operations" (CAFO), or "intensive livestock operations", can hold large numbers (some up to hundreds of thousands) of cows, hogs, turkeys, or chickens, often indoors. The essence of such farms is the concentration of livestock in a given space. The aim is to provide maximum output at the lowest possible cost and with the greatest level of food safety. The term is often used pejoratively. However, CAFOs have dramatically increased the production of food from animal husbandry worldwide, both in terms of total food produced and efficiency.

Food and water is delivered to the animals, and therapeutic use of antimicrobial agents, vitamin supplements, and growth hormones are often employed. Growth hormones are not used on chickens nor on any animal in the European Union. Undesirable behaviors often related to the stress of confinement led to a search for docile breeds (e.g., with natural dominant behaviors bred out), physical restraints to stop interaction, such as individual cages for chickens, or physical modification such as the de-beaking of chickens to reduce the harm of fighting.

The CAFO designation resulted from the 1972 U.S. Federal Clean Water Act, which was enacted to protect and restore lakes and rivers to a "fishable, swimmable" quality. The United States Environmental Protection Agency (EPA) identified certain animal feeding operations, along with many other types of industry, as "point source" groundwater polluters. These operations were subjected to regulation.

In 17 states in the U.S., isolated cases of groundwater contamination were linked to CAFOs. For example, the ten million hogs in North Carolina generate

19 million tons of waste per year. The U.S. federal government acknowledges the waste disposal issue and requires that animal waste be stored in lagoons. These lagoons can be as large as 7.5 acres (30,000 m). Lagoons not protected with an impermeable liner can leak into groundwater under some conditions, as can runoff from manure used as fertilizer. A lagoon that burst in 1995 released 25 million gallons of nitrous sludge in North Carolina's New River. The spill allegedly killed eight to ten million fish.

The large concentration of animals, animal waste, and dead animals in a small space poses ethical issues to some consumers. Animal rights and animal welfare activists have charged that intensive animal rearing is cruel to animals.

Other concerns include persistent noxious odor, the effects on human health, and the role of antibiotic use in the rise of resistant infectious bacteria.

According to the U.S. Centers for Disease Control and Prevention (CDC), farms on which animals are intensively reared can cause adverse health reactions in farm workers. Workers may develop acute and/or chronic lung disease, musculoskeletal injuries, and may catch (zoonotic) infections from the animals.

Managed intensive rotational grazing

Managed Intensive Rotational Grazing (MIRG), also known as cell grazing, mob grazing, and holistic management planned grazing, is a variety of foraging in which herds or flocks are regularly and systematically moved to fresh, rested grazing areas to maximize the quality and quantity of forage growth. MIRG can be used with cattle, sheep, goats, pigs, chickens, turkeys, ducks, and other animals. The herds graze one portion of pasture, or a paddock, while allowing the others to recover. Resting grazed lands allows the vegetation to renew energy reserves, rebuild shoot systems, and deepen root systems, resulting in long-term maximum biomass production. MIRG is especially effective because grazers thrive on the more tender younger plant stems. MIRG also leaves parasites behind to die off, minimizing or eliminating the need for de-wormers. Pasture systems alone can allow grazers to meet their energy requirements, and with the increased productivity of MIRG systems, the animals obtain the majority of their nutritional needs, in some cases all, without the supplemental feed sources that are required in continuous grazing systems.

Pasture intensification

Pasture intensification is the improvement of pasture soils and grasses to increase the food production potential of livestock systems. It is commonly used to reverse pasture degradation, a process characterized by loss of forage and decreased animal carrying capacity which results from overgrazing, poor nutrient management, and lack of soil conservation. This degradation leads to poor pasture soils with decreased fertility and water availability and increased rates of erosion, compaction, and acidification. Degraded pastures have significantly lower productivity and higher carbon footprints compared to intensified pastures.

Management practices which improve soil health and consequently grass productivity include irrigation, soil scarification, and the application of lime, fertilizers, and pesticides. Depending on the productivity goals of the target agricultural system, more involved restoration projects can be undertaken to replace invasive and under-productive grasses with grass species that are better suited to the soil and climate conditions of the region. These intensified grass systems allow higher stocking rates with faster animal weight gain and reduced time to slaughter, resulting in more productive, carbon-efficient livestock systems.

Another technique to optimize yield while maintaining the carbon balance is the use of integrated crop-livestock (ICL) and crop-livestock-forestry (ICLF) systems, which combine several ecosystems into one optimized agricultural framework. These synergies between these systems provide benefits to pastures through optimal plant usage, improved feedand fattening rates, increased soil fertility and quality, intensified nutrient cycling, integrated pest control, and improved biodiversity. The introduction of certain legume crops to pastures increases carbon accumulation and nitrogen fixation in soils, while their digestibility helps animal fattening and reduces methane emissions from enteric fermentation.ICLF systems yield beef cattle productivity up to ten times that of degraded pastures, additional crop production from maize, sorghum, and soybean harvests, and greatly reduced greenhouse gas balances due to forest carbon sequestration.

Crops

The Green Revolution transformed farming in many developing countries. It spread technologies that had already existed, but had not been widely used outside of industrialized nations. These technologies included "miracle seeds", pesticides, irrigation, and synthetic nitrogen fertilizer.

Seeds

In the 1970s, scientists created strains of maize, wheat, and rice that are generally referred to as high-yielding varieties (HYV). HYVs have an increased nitrogen-absorbing potential compared to other varieties. Since cereals that absorbed extra nitrogen would typically lodge (fall over) before harvest, semidwarfing genes were bred into their genomes. Norin 10 wheat, a variety developed by Orville Vogel from Japanese dwarf wheat varieties, was instrumental in developing wheat cultivars. IR8, the first widely implemented HYV rice to be developed by the International Rice Research Institute, was created through a cross between an Indonesian variety named "Peta" and a Chinese variety named "Dee Geo Woo Gen."

With the availability of molecular genetics in Arabidopsis and rice the mutant genes responsible (reduced height (rht), gibberellin insensitive (gai1) and slender rice (slr1)) have been cloned and identified as cellular signalling components of gibberellic acid, a phytohormone involved in regulating stem growth via its effect on cell division. Photosynthetic investment in the stem is reduced dramatically as the shorter plants are inherently more mechanically stable. Nutrients become redirected to grain production, amplifying in particular the yield effect of chemical fertilizers.

HYVs significantly outperform traditional varieties in the presence of adequate irrigation, pesticides, and fertilizers. In the absence of these inputs, traditional varieties may outperform HYVs. They were developed as F1 hybrids, meaning seeds need to be purchased every season to obtain maximum benefit, thus increasing costs.

Crop rotation

Crop rotation or crop sequencing is the practice of growing a series of dissimilar types of crops in the same space in sequential seasons for benefits such as avoiding pathogen and pest buildup that occurs when one species is continuously cropped. Crop rotation also seeks to balance the nutrient demands of various crops to avoid soil nutrient depletion. A traditional component of crop rotation is the replenishment of nitrogen through the use of legumes and green manure in sequence with cereals and other crops. Crop rotation can also improve soil structure and fertility by alternating deep-rooted and shallow-rooted plants. A related technique is to plant multi-species cover crops between commercial crops. This combines the advantages of intensive farming with continuous cover and polyculture.

Irrigation

Crop irrigation accounts for 70% of the world's fresh water use. Flood irrigation, the oldest and most common type, is typically unevenly distributed, as parts of a field may receive excess water in order to deliver sufficient quantities to other parts. Overhead irrigation, using center-pivot or lateral-moving sprinklers, gives a much more equal and controlled distribution pattern. Drip irrigation is the most expensive and least-used type, but delivers water to plant roots with minimal losses.

Water catchment management measures include recharge pits, which capture rainwater and runoff and use it to recharge groundwater supplies. This helps in the replenishment of groundwater wells and eventually reduces soil erosion. Dammed rivers creating reservoirs store water for irrigation and other uses over large areas. Smaller areas sometimes use irrigation ponds or groundwater.

Weed control

In agriculture, systematic weed management is usually required, often performed by machines such as cultivators or liquid herbicide sprayers. Herbicides kill specific targets while leaving the crop relatively unharmed. Some of these act by interfering with the growth of the weed and are often based on plant hormones. Weed control through herbicide is made more difficult when the weeds become resistant to the herbicide. Solutions include:

- Cover crops (especially those with allelopathic properties) that out-compete weeds or inhibit their regeneration
- Multiple herbicides, in combination or in rotation
- Strains genetically engineered for herbicide tolerance
- Locally adapted strains that tolerate or out-compete weeds
- Tilling
- Ground cover such as mulch or plastic
- Manual removal
- Mowing
- Grazing
- Burning

Terracing

In agriculture, a terrace is a leveled section of a hilly cultivated area, designed as a method of soil conservation to slow or prevent the rapid surface runoff of irrigation water. Often such land is formed into multiple terraces, giving a stepped appearance. The human landscapes of rice cultivation in terraces that follow the natural contours of the escarpments, like contour ploughing, are a classic feature of the island of Bali and the Banaue Rice Terraces in Banaue, Ifugao, Philippines.

In Peru, the Inca made use of otherwise unusable slopes by building drystone walls to create terraces.

Rice paddies

A paddy field is a flooded parcel of arable land used for growing rice and other semiaquatic crops. Paddy fields are a typical feature of rice-growing countries of east and southeast Asia, including Malaysia, China, Sri Lanka, Myanmar, Thailand, Korea, Japan, Vietnam, Taiwan, Indonesia, India, and the Philippines. They are also found in other rice-growing regions such as Piedmont (Italy), the Camargue (France), and the Artibonite Valley (Haiti). They can occur naturally along rivers or marshes, or can be constructed, even on hillsides. They require large water quantities for irrigation, much of it from flooding. It gives an environment favourable to the strain of rice being grown, and is hostile to many species of weeds. As the only draft animal species which is comfortable in wetlands, the water buffalo is in widespread use in Asian rice paddies.

Paddy-based rice-farming has been practiced in Korea since ancient times. A pit-house at the Daecheon-ni archaeological site yielded carbonized rice grains and radiocarbon dates indicating that rice cultivation may have begun as early as the Middle Jeulmun Pottery Period (c. 3500-2000 BC) in the Korean Peninsula. The earliest rice cultivation there may have used dry-fields instead of paddies.

The earliest Mumun features were usually in naturally swampy, low-lying narrow gulleys and fed by local streams. Some Mumun paddies in flat areas were made of a series of squares and rectangles separated by bunds approximately 10 cm in height, while terraced paddies were long and irregular in shape, following the natural contours of the land at various levels.

Like today, Mumun period rice farmers used terracing, bunds, canals, and small reservoirs. Some paddy-farming techniques of the Middle Mumun period (c. 850-550 BC) can be interpreted from the well-preserved wooden tools excavated from archaeological rice paddies at the Majeon-ni site. Iron tools for paddy-farming were not introduced until sometime after 200 BC. The spatial scale of individual paddies, and thus entire paddy-fields, increased with the regular use of iron tools in the Three Kingdoms of Korea Period (c.AD 300/400-668). A recent development in the intensive production of rice is System of Rice Intensification (SRI). Developed in 1983 by the French Jesuit Father Henri de Laulanié in Madagascar,by 2013 the number of smallholder farmers using SRI had grown to between 4 and 5 million.

Aquaculture

Aquaculture is the cultivation of the natural products of water (fish, shellfish, algae, seaweed, and other aquatic organisms). Intensive aquaculture takes place on land using tanks, ponds, or other controlled systems, or in the ocean, using cages.

SUSTAINABLE INTENSIVE FARMING

Sustainable intensive farming practices have been developed o slow the deterioration of agricultural land and even regenerate soil health and ecosystem services, while still offering high yields. Most of these developments fall in the category of organic farming, or the integration of organic and conventional agriculture. "Organic systems and the practices that make them effective are being picked up more and more by conventional agriculture and will become the foundation for future farming systems. They won't be called organic, because they'll still use some chemicals and still use some fertilizers, but they'll function much more like today's organic systems than today's conventional systems."

Dr. Charles Benbrook Executive director US House Agriculture Subcommittee Director Agricultural Board - National Academy Sciences (FMR) The System of Crop Intensification (SCI) was born out of research primarily at Cornell University and smallholder farms in India on SRI. It uses the SRI concepts and methods for rice and applies them to crops like wheat, sugarcane, finger millet, and others. It can be 100% organic, or integrated with reduced conventional inputs.

Holistic management is a systems thinking approach that was originally developed for reversing desertification. Holistic planned grazing is similar to rotational grazing but differs in that it more explicitly provides a framework for adapting to four basic ecosystem processes: the water cycle, the mineral cycle (including the carbon cycle), energy flow, and community dynamics (the relationship between organisms in an ecosystem) as equal in importance to livestock production and social welfare. By intensively managing the behavior and movement of livestock, holistic planned grazing simultaneously increases stocking rates and restores grazing land.

Pasture cropping involves planting grain crops directly into grassland without first applying herbicides. The perennial grasses form a living mulch understory to the grain crop, eliminating the need to plant cover crops after harvest. The pasture is intensively grazed both before and after grain production using holistic planned grazing. This intensive system yields equivalent farmer profits (partly from increased livestock forage) while building new topsoil and sequestering up to 33 tons of CO2/ha/year.

The Twelve Aprils grazing program for dairy production, developed in partnership with the USDA-SARE, is similar to pasture cropping, but the crops planted into the perennial pasture are forage crops for dairy herds. This system improves milk production and is more sustainable than confinement dairy production. Integrated multi-trophic aquaculture (IMTA) is an example of a holistic approach. IMTA is a practice in which the by-products (wastes) from one species are recycled to become inputs (fertilizers, food) for another. Fed aquaculture (e.g. fish, shrimp) is combined with inorganic extractive (e.g. seaweed) and organic extractive aquaculture (e.g. shellfish) to create balanced systems for environmental sustainability (biomitigation), economic stability (product diversification and risk reduction), and social acceptability (better management practices).

Biointensive agriculture focuses on maximizing efficiency such as per unit area, energy input, and water input. Agroforestry combines agriculture and orchard/forestry technologies to create more integrated, diverse, productive, profitable, healthy, and sustainable land-use systems.

Intercropping can increase yields or reduce inputs and thus represents (potentially sustainable) agricultural intensification. However, while total yield per acre is often increased dramatically, yields of any single crop often diminish. There are also challenges to farmers relying on farming equipment optimized for monoculture, often resulting in increased labor inputs.

Vertical farming is intensive crop production on a large scale in urban centers, in multi-story, artificially-lit structures, using far less inputs and producing fewer environmental impacts.

An integrated farming system is a progressive, biologically-integrated sustainable agriculture system such as IMTA or Zero waste agriculture, whose implementation requires exacting knowledge of the interactions of multiple species and whose benefits include sustainability and increased profitability. Elements of this integration can include:

- Intentionally introducing flowering plants into agricultural ecosystems to increase pollen-and nectar-resources required by natural enemies of insect pests
- Using crop rotation and cover crops to suppress nematodes in potatoes

Challenges

The challenges and issues of industrial agriculture for society, for the industrial agriculture sector, for the individual farm, and for animal rights include the costs and benefits of both current practices and proposed changes to those practices. This is a continuation of thousands of years of invention in feeding ever-growing populations.

[W]hen hunter-gatherers with growing populations depleted the stocks of game and wild foods across the Near East, they were forced to introduce agriculture. But agriculture brought much longer hours of work and a less rich diet than hunter-gatherers enjoyed. Further population growth among shifting slash-and-burn farmers led to shorter fallow periods, falling yields and soil erosion. Plowing and fertilizers were introduced to deal with these problems - but once again involved longer hours of work and degradation of soil resources (Boserup, The Conditions of Agricultural Growth, Allen and Unwin, 1965, expanded and updated in Population and Technology, Blackwell, 1980.).

While the point of industrial agriculture is to profitably supply the world at the lowest cost, industrial methods have significant side effects. Further, industrial agriculture is not an indivisible whole, but instead is composed of multiple elements, each of which can be modified in response to market conditions, government regulation, and further innovation, and has its own side-effects. Various interest groups reach different conclusions on the subject.

Population growth

- 30,000 years ago hunter-gatherer behavior fed 6 million people
- 3,000 years ago primitive agriculture fed 60 million people
- 300 years ago intensive agriculture fed 600 million people
- Today industrial agriculture attempts to feed 8 billion people

Between 1930 and 2000, U.S. agricultural productivity (output divided by all inputs) rose by an average of about 2 percent annually, causing food prices to decrease. "The percentage of U.S. disposable income spent on food prepared at home decreased, from 22 percent as late as 1950 to 7 percent by the end of the century."

Other impacts

Industrial agriculture uses huge amounts of water, energy, and industrial chemicals, increasing pollution in the arable land, usable water, and atmosphere.

Herbicides, insecticides, and fertilizers are accumulating in ground and surface waters. "Many of the negative effects of industrial agriculture are remote from fields and farms. Nitrogen compounds from the Midwest, for example, travel down the Mississippi to degrade coastal fisheries in the Gulf of Mexico. But other adverse effects are showing up within agricultural production systems-for example, the rapidly developing resistance among pests is rendering our arsenal of herbicides and insecticides increasingly ineffective."Agrochemicals and monoculture have been implicated in Colony Collapse Disorder, in which the individual members of bee colonies disappear. Agricultural production is highly dependent on bees to pollinate many varieties of fruits and vegetables.

Social

A study done for the U.S. Office of Technology Assessment conducted by the UC Davis Macrosocial Accounting Project concluded that industrial agriculture is associated with substantial deterioration of human living conditions in nearby rural communities.

Land banking

Land banking is the practice of aggregating parcels of land for future sale or development. While in many countries land banking may refer to various private real-estate investment schemes, in the United States it refers to the establishment of quasi-governmental county or municipal authorities purposed with managing an inventory of surplus land.

Municipal land banks in the United States

Land banks are quasi-governmental entities created by counties or municipalities to effectively manage and repurpose an inventory of underused, abandoned, or foreclosed property. They are often chartered to have powers that enable them to accomplish these goals in ways that existing government agencies can not. While the land bank "model" has gained broad support and has been implemented in a number of cities, they are implemented differently so as to best address both municipal needs and the state and local legal context in which they were created.

History

The period of deindustrialization in the United States coupled with increased suburbanization in the middle of the 20th century left many American cities with large amounts of vacant & blighted industrial, residential, and commercial property. Beginning in the early 1970s municipalities began to seek solutions to manage decline or spur revitalization in once prosperous city neighborhoods. The first land bank was created in St.

Louis in 1971. While additional municipalities continued to adopt them at a trickle it wasn't until the mid 2000s that land banks became viewed as a tested, reliable, and accepted model and experienced widespread implementation - particularly after the success of the Genesee County Land Bank. In 2009 the Department of Housing and Urban Development issued a report embracing land banks as a best practices model for municipalities dealing with the effects of the real estate market collapse and the ensuing foreclosure crisis.

INVESTMENT LAND BANKING BY COUNTRY

Land banking in the UK was previously the preserve of the landed gentry or real estate developers. Many reputable listed commercial building companies engage successfully in land banking for future building projects. Companies also purchase land sites and easily divide them into smaller plots, then offer these plots for sale to individual investors. This relatively new practice in the UK does not fall under the control of the Financial Services Authority. Many people are wary of this form of investment as many plot based land banking companies have failed or been closed down. There are currently no audited successes recorded for UK plot based land banking despite the UK having gone through a major property boom in the period 2002 - Jul 2007.

A landbanking scheme that is a Collective investment scheme is a "regulated activity" for the purposes of the Financial Services and Markets Act 2000 and, according to section 19(1), may only be operated in the UK by a person who is either authorised or exempt. Section 26 provides that an agreement made by a person in contravention of this is unenforceable and any sums paid to him may be recovered together with compensation for any loss suffered. After recent FSA enforcement of this regulation many companies selling UK land plots have moved outside of the European Union and only offer land plots to non UK residents who are not protected by FSA regulations.

Companies offering land banking plots in the UK

Since the changes in the land registration act, a number of companies offering UK land plots as an investment have been formed. Typically this land is greenbelt, nature conservation, flood plain, agricultural or protected land unsuitable for development. There are no recorded successful planning permission applications for plots sold under such collective investment schemes.

There have been considerable losses recorded by investors in UK land plot investment schemes. A large number of British companies offering UK land plots have failed or been shut down by the FSA or other authorities. Some companies have now moved offshore after FSA investigation. Some companies now offer UK land plots from locations such as Dubai or Singapore where the local authorities do not regulate such activities, or are not aware of the high risk nature of the investment. In June 2010 the Monetary Authority of Singapore (MAS) issued a warning on land banking plots schemes warning they may be a scam with specific focus on companies offering land from the UK and Canada.

Sales methods

A company representative may contact an individual by telephone, in temporary shopping center booths, or at property shows and offer a strategic land investment in the UK. Very often UK government or industry statistics, the proximity of the land to built up areas, or the recent history of UK house prices are quoted as a demonstration of why the land plot is a great investment. Verbal communication will often indicate that the land is fast tracked for building approval and has strong potential as building land. When pricing the land reference is typically made to approved building land prices at the market peak. Very often the land banking company will present detailed plans showing a housing development on the site. These plans are often referred to as "pre-approved", "concept" or "predevelopment". The sales person will focus on the potential future value of the land against the current selling price.

No reference is ever made to the value of green belt or agricultural land, or the issues involved with the long term maintenance, and collectively selling tiny plots of land. The sales price is typically uplifted 10-100 times over the current value of the land. Plans shown have no validity in UK planning law and cannot be considered as any indication of progress in the planning process. No written contractual promise is ever given for planning permission despite typically extreme optimism by the sales person. The sales person will typically never mention that the land is protected, or greenbelt land and cannot be developed under current planning regulations. There is typically no possibility of planning permission in any reasonable timeframe.

The investor may end up paying a considerable amount of money for a small area of low value land which has a very high risk of standing undeveloped. Once the general public becomes aware of the lack of viability of the proposed plot investment scheme, the re-sale value of the individual plots collapses. This is typically followed by the land plot company liquidating completely, or relocating to another legal jurisdiction.

For customers that show a willingness to purchase such schemes there may also be attempts to sell additional plot based land banking products at alternate locations, or other high yield investment programmes. Customers may also be added to suckers lists which are then sold to other companies offering similar schemes. When the land banking plot company fails plot investors may also be offered investment recovery or planning services for a fee. Such services typically are fraudulent or fail and lead to a further loss of money for the investor.

Controversies

A You and Yours documentary, first aired on BBC Radio 4 in December 2006, criticized the services offered by many land banking companies in the United Kingdom, suggesting that they were scamming their customers. A land banking scam is based on the very low chance of any of the plots receiving planning permission and the very high profit margins taken on the land plots, with the seller using misleading marketing tactics to convince the buyers that they are making a sound investment.

A key strategy used for selling United Kingdom land plots is to imply that because a customer owns the land plot, they cannot lose their money. The land banking company typically suggests dramatic annual increases in the value of the land plots, and a very optimistic time frame for successful planning applications. These are never contractually committed. Typically the land banking company sells a land plot at a premium of 15 to 100 times the current market value of undeveloped land. A purchaser might pay £15,000 for a land plot that only has a current market value of £500. On this basis, most of the investment is not in land, and the small percentage annual increases in the value of the land plot are meaningless. The actual investment is in a proposed service to deliver valuable approved building land in the future. If that service is never delivered or is not successful, the remaining land asset is normally worthless. Should the selling company fail or disappear, the plot owner cannot economically sell the plot, as the administrative effort and cost of sale typically exceeds the value of the land plot.

Many land banking companies target victims outside of the United Kingdom, particularly Canada, Singapore, Thailand, Brunei and Malaysia. Residents of these countries may be naive of the UK property market and local planning regulations such as green belt zoning.

In 2008, the land banking firm UKLI was placed into administration due to insolvency, despite having taken £69 million from 4,500 people for land plots.

Land International was closed down in 2008 after losing investors £10 million, and the same Land International plots were later offered for sale in Asia. In 2010, Land International (Far East) failed, causing investors to lose S$6 million (£2.5M). MP David Heath requested a debate in the House of Commons following the offering of 209 plots in the village of Dean, saying that "while land banking may not be illegal it is undoubtedly a scam".

The UK Land Registry issued a press release on January 15, 2009 advising consumers that the Land Registry has published a guide warning against land banking investment schemes. Land Registry Head of Corporate Legal Services Mike Westcott Rudd said that the public were being "misled about the prospects of obtaining planning permission", with well-known banks and developers being falsely cited as partners in the project, and that in some cases forged Land Registry paperwork was being presented to suggest that planning approval existed where it did not.

United States

In 2011 New York State passed a land bank statute authorizing the establishment of nonprofits in each county to take title to vacant abandoned homes so they can be rehabilitated, sold or demolished in an orderly fashion. Many counties upstate including Erie County, Onondaga County, Schenectady County and Albany County have abandoned homes as people moved to the suburbs. Some properties have been abandoned due to back taxes and the city has taken title. The recent robo-signing settlement gave Attorney General Eric Schneiderman the wherewithal to fund land banks in Schenectady and Albany. The state of Michigan also has a land bank program. Ohio passed land bank legislation in 2009

3

HORTICULTURAL CROP GROWERS

Changes in Socio-Economic structure are very closely related to change in economic development. This is true both at the macro as well as at the micro level. This profile of producers have a profound influence on the decision making process and profitability of crop enterprise. An attempt has been made in this chapter to compare and contrast the resource endowments of different categories of farms in different states or different export oriented horticultural crops growing area. The information relating to family size and composition, education, land use pattern, cropping pattern, source of irrigation, farm implements and machinery and livestock resources at various farms in the different selected state have been analysed and discussed for various categories of orchardist/growers. The knowledge of the background of the sample producers is essential since the viability of any enterprise heavily depends on the favourable attitudinal changes towards adoption of superior technical inputs or technique of production which, in turn, depends on technical skills and resource position of growers.

The socio-economic resource profile of sample growers in different states are as follow:-

ANDHRA PRADESH

Out of the total population of the sample 268, males constitute 49.25 percent, while females account for the remaining 50.75 per cent in dictating a modest female predominance among the sample households. Among the size groups, while female predominance is observed in the marginal and medium sizegroup of farmers, males had an edge over female population among the small farmers. However equal number is maintained in the case of large farmers. The rate of literacy is as high as 79.48 percent, among the sample households and a majority of them are at secondary or higher education. Female literacy is also considerably high constituting about 47.89 percent of the total literates. This is mainly due to the proximity of the sample villages to the urban centre of Hyderabad and most of the sample households belonged to the socially and economically advanced communities. Rate of literacy and educational status of the sample households.

The area operated by the sample households is completely owned by them and no leasing of land is reported in the sample villages. The per holding area operated by the sample households is 2.82 hectares and the net area sown is 2.81 hectares with 0.01 hectares as current fallow. The entire land owned and operated by the sample households is irrigated and is under orchards except a small extent under paddy. Since grape cultivation is highly capital and labour intensive, the sample farmers have concentrated only on this crop as 97.69 per cent of the total cropped area of 140.44 hectares is under grapes. Paddy is the only other crop grown on a small extent and its share in the total cropped area is only 2.31 per cent.

Since grape is a commercial crop and the land is well irrigated, value of land formed a major proportion of 65.59% of the total per household value of assets owned by the sample households. Residential buildings shared 20.46% of the total value of assets as most of the cultivators are living in semi urban areas, some of them in posh buildings. The value of farm implements and machinery like electric motors, oil engines, pumpsets, tractors, sprayers and dusters constituted about 9.85% of the total value of assets. While the value of cattlesheds, bullock carts and farm houses constituted 2.26% of the total value of assets, value of live stock accounted for only 0.88% as most of the farm operations are conducted by tractors and other modern equipment..

BIHAR

The socio-economic profile explains that all the villages of the study area are well connected by both road and railway which is a boon for the transportation of fruit and other commodities produced in the project area. The soil of the area is calcareous containing free calcium carbonate to the tune of 21 per cent to 40 per cent which is relatively beneficial for the production of litchi. The soil reaction is alkaline and the PH ranges between 8.3 to 8.7. The climate and soil of all the selected villages were identical. The area has humid subtropical climate with or summer, moderate rainfall and cold winter. The demographic features of the sample orchardist revealed that out of 50 respondents 8.0 per cent belong to scheduled castes, 38.0 per cent backward castes and 54.0 per cent to general caste. The average size of family is 6.9 persons per household which ranged from 9.25

persons on marginal category to 5.75 persons on medium category of orchardists. The 99.50 per cent respondent were literate and they were mainly engaged in agriculture (54.0%). In all the three villages have assured Irrigational facilities mainly in the form of diesel pumpsets which was evident from the fact that the net irrigated area constituted nearly 73.96, 72.31 and 72.11 per cent of net sown area of the village Kothia, kaparpura and Kalwari respectively. The cropping pattern of the selected villages shows orchard dominated. About 50 per cent area in the cropping pattern of the selected villages were covered with orchard, moreover, litchi orchard covers about 35 per cent area in the total gross cropped area. The orchardist of the project area were well equipped with farm implements and they were rearing animals for dairy, poultry, draft, etc. purposes.

GUJARAT

Majority of households i.e. 58 per cent belonged to brahmin caste known as Anavils and 16 per cent to schedule tribe. 93.3 per cent of total population was literate. 38 per cent of head of the households had attained education upto graduate level, while 40 per cent had attained education upto secondary school. About 66 per cent of the total population was economically active.

Overall per hectare investment on all the assets was Rs. 47300, of which 58.6 per cent was for tractors, 11.4 per cent for tubewells, 8.7 per cent for wells and 4.0 per cent for threshers.

On the whole, 74 per cent of sample households had owned cows, 38 per cent had buffaloes and 4 per cent had owned bullocks.

Overall 2.6 hectares per household was owned land and 0.1 hectare was leased-in land. The share of waste land was negligible. There was complete absence of mortgage transactions. Majority area of sample households was dominated by orchards of chikus and mangoes. Low acreage was spared for field crops. The area under chiku was positively related with size of land holding group.

The GCA of each sample household was under irrigation. Wells, tubewells and canals were the major sources of irrigation.

HARYANA

The commercial flower cultivation can be likened to an industrial production process and therefore has generally been practiced by a handful of big or well off farmers or by a few resourceful entrepreneurs.

Given the nature of the crop selected for study, its special cultivation requirements and characteristics, the fact that its cultivation has started only on a limited scale and that too only recently, the selected crop is not ammenable to detailed investigation in respect of some of the issues stipulated by the study coordinator for horticultural crops in general due to non-availability of required data on a number of parameters. Further our repeated attempts to interview the owners/promoters and/or their managers of these commercial floriculture units to ascertain information on some of the aspects relevant for the study were usually not met with much success. Most of the entrepreneurs, it was felt, did not want to share their experiences specially about cost of production and returns etc. for obvious reasons. Accordingly the objectives of the present study have been somewhat modified keeping in view the specificity of the crop under study, the prevailing production environment and the data availability constraints.

HIMACHAL PRADESH

Out of 126 sampled orchardists 50, 43 and 33 are from Shimla, Kullu and Mandi districts respectively. The second largest group of sample orchardists is of small category. Out of total sample orchardists more than 15 per cent belong to scheduled caste category. The average family size varies from 5.3 persons on marginal category to 7.8 persons on medium size of orchardists and the family size is almost the same in all the selected districts. Among different districts the number of labour force is almost same and average being 3.4 persons. The literacy percentage is reported more among male members on different categories of orchardists and almost same for male and female members in Mandi district. The literacy percentage of females is lower (41.3%) in

Shimla district as compared to the State average of 52 per cent (Census 1991). The occupational distribution reveals that more than 68 per cent male workers are engaged in agriculture while the percentage of female workers account for 80.3 per cent.

The land holding size of sample orchardists varies from 0.58 hectare on marginal farm to 5.28 hectares on large farm, the average being 1.28 hectares. Out of total land owned by sample orchardists about 6 per cent area is reported irrigated on average farm and this proportion is reported higher (13.27%) on medium farms. In Shimla and Mandi districts no area is reported irrigated. In Kullu district the proportion of irrigated land is more than 35 per cent. In Mandi district no field crop was grown by sample orchardists except inter-cropping in young orchards, while in Shimla district no field crop was grown by the sample orchardist. In Kullu district average under pure field crops is more than 32 per cent.

MAHARASHTRA

The average size of a family consisted of 6 persons in the case of onion producers and nearly 8 persons in the case of grape orchardists. The family size of large category was seen to be more compared to other categories, both in the case of onion producers and grape orchardists. A positive association was seen between land holding size and number of members in the family, that is, the number of members in the family increased from marginal to small, small to medium and medium to large category. This held true for both onion producers and grape orchardists. In order to ascertain the extent of education of the members of a family, an index of literacy was computed for various categories of onion producers and grape orchardist. The overall literacy index was seen to be slightly higher for onion producers compared to grape orchardists. In case of onion producers, the literacy index was seen to increase with the increase in size of land holding. On an average, 0.15 acre of land was seen to be acquired on lease by small category of onion producers. The intensity of cropping was ranged from 147 per cent in case of marginal category of onion producers to 126 per cent for large category with an average of 139 per cent for the average category of onion producers. The area under kharif crops in case of grape orchardists varied from 0.75 acre on marginal farms to 7.50 acres on large farms. In case of rabi crops, it varied from 0.75 acre on marginal farms to 3.73 acres on large farms. In case of sample onion producers, the crops other than onion that dominated the cropping pattern of these producers were seen to be hybrid jowar, bajra, groundnut and soyabean in kharif season and wheat and hybrid gram in rabi season. Grapes, sugarcane and Lucerne were seen to be grown as perennial crops by these sample onion producers.

In general, in case of perennial crops, 80 per cent of the total cropped area was seen to be under grape orchards, 16 per cent under sugarcane, 1 per cent under Lucerne and the remaining 3 per cent under pomegranate. Groundwater based irrigation is seen to have dominated over the surface water based irrigation in the study area. In general 90 per cent of the net irrigated area of sample onion producers was under well irrigation, about 5 per cent under canal irrigation and the remaining 5 per cent under other sources of irrigation. In onion producer, milch cows accounted for 32 per cent, milch buffaloes 25 per cent, bullocks 12 per cent, he-buffaloes 9 per cent and young stock 20 per cent. The sample grape orchardists possessed less number of animals compared to sample onion producers. On an average, the grape orchardists possessed around 4 animals in standard animal unit.

SIKKIM

Regarding the general features of the sample orchid growers, it is found that more than 26 per cent of the orchid growers are marginal farmers and 41.25 per cent are small farmers. The percentage of medium and large farmers in the sample are 15.00 per cent and 17.50 per cent respectively. In terms of number of pots of orchid, majority of the growers (36.25 per cent) fall into the category of growers who hold 101 - 200 pots. The potwise composition of other farms are 28.75 per cent in the category of farms having less than 100 pots, 20 per cent in the category having 201-300 pots and rest 15.00 per cent in the category having 301 & above pots.

As per the location feature of the sample farms, it is observed that more than 56 per cent farms are located within a distance upto 1 km from the permanent road and 28.75 per cent of the farms are located within a distance from 1.01-2 kms. The rest of the sample farms are located within a distance from 2.01-3kms. (8.75 per cent) and 3 kms. and above (6.25 per cent).

Out of the eighty growers, more than 21 per cent belongs to the scheduled caste and 10 per cent belongs to the scheduled tribe. It is also fact that the scheduled caste and scheduled tribe growers are mainly marginal and small farmers. The average size of family of the sample orchid growers varies from 5.4 in medium category to 6.5 in marginal category. The average family size in small and large categories of farms are 6.5 and 6.0 respectively. it is also found that proportion of male worker is more in medium category farms followed by large and small categories and it is lowest in marginal farms. Similarly, the average work force is found to be more in medium and large farms than that of marginal and small farms. The literacy percentage is also less in marginal and small categories of growers than that of large farms. The overall literacy percentage is found to be 55.88 per cent. The land holding size of the sample farms varies from 2.21 acres in marginal farms to 12.52 acres in large farms. The overall farm size is found to be 5.85 acres. Among the field crops grown by the orchid growers, maize accounts for the major share (39.23 per cent) and the share of maize in the cropping pattern is found to be highest in large farms (41.53 per cent). The cropping pattern of these farms other than maize is concentrated towards cultivation of other cereals like paddy and wheat and vegetables like cabbage, cauliflower, potato and cardamom. But the share of vegetables in the cropping pattern varies from farm to farm which may be due to difference in Irrigational facilities and altitude of land. However, there is no evidence that flowers are grown by the growers other than orchid by the sample farms.

TRIPURA

There are altogether 320 persons in the 50 sample households. The average size of family is 6.4. The proportion of male and female in the population is 52.50 and 47.50 respectively. Total number of persons in the broad age-group of 15 years to 65 years is 217. This 67.81% people constitute the working force of the sample households. Of course, some others from below 15 years of age and above 65 years also do participate in agricultural operations and other household works. The number of female per 1000 male is worked out to be 905. The rate of literacy for the sample population as a whole is 75.63 per cent while the rate of literacy among the males is 81.55 per cent and for females it is 69.08 per cent. Regarding the land resources the total geographical area is 10,492 sq.km. out of which about 60 per cent are covered by hills and forest and only 40 per cent is considered as plains. The cropping pattern of the sample farmers is very simple. The farmers have two types of land viz. plain land and tilla land. The former is used for field crop while the latter for horticultural crops. The main livestock reared by the sample households are cows, pigs and poultry birds. The 50 sample households possess only 24 bullocks which are used as draught animal. Possession of small area under field crops (like paddy) in each household leads to disproportionate number of bullocks among the sample households.

UTTER PRADESH

Distribution of total family members on the farms of sample mango growers in Saharanpur district indicates that on an average the total number of family members per sample mango growers was 10 of which 3 were males, 3 females and remaining 4 were children. Whereas the total number of family members per farm in case of potato growers in Farrukhabad district was 9.62 on an average of which 3 were males, 2.46 were females and 4.16 children. Land utilization on the sample farms of mango growers in Saharanpur district indicates that the total operated area per farm was estimated to 5.06 hectare, on an overall average. Among the various categories of mango growers, the operated area was found to be highest i.e., 9.90 hectare per farm in the category of bearing stage having decreasing production against the lowest i.e., 1.96 hectare per farm in the category of non-bearing stage of mango. the area per farm was found increasing in various categories with the increase in the age of mango. On the other hand land utilization on the sample farms of potato growers in Farrukhabad district indicates that the total operated area per farm was estimated to 2.32 hect. on an overall average. While the net cultivated area per farm was 2.35 hectare of the total owned area of 2.38 hectare. Cropping pattern followed on the sample farms of mango growers in Saharanpur district indicates that on an overall the total area covered by perennial fruit crops was 3.85 hectares per farm. While the average per farm in the category of non bearing stage under perennial fruit crops was 1.14 hectare. Cropping pattern followed on the sample farms of potato growers in Farrukhabad district indicates that on the overall average the gross cropped area per farm was estimated to 6.04 hectares. While the highest gross cropped area per farm was found on the farms of large potato growers against the lowest i.e. 1.47 hectares per farm in the category of marginal potato growers. Distribution of farm inventories on the sample farms of mango growers in Saharanpur district indicates that the value of total farm inventories on an average was estimated to Rs. 9,01,775 per farm. The distribution of farm inventories on the sample farms of potato growers in Farrukhabad district indicates that on an overall average the total value of farm inventories was accounted Rs.9,01,395 per farm.

WEST BENGAL

Regarding the general features of the sample orchid growers, it is found that the more than 32 per cent of the orchid growers are marginal farmers and 40 per cent are small farmers. The percentage of medium and large farmers in the sample are 13.75 per cent each. In terms of number of pots of orchid, majority of the growers (40 per cent) fall into the category of growers holding upto 100 pots. The potwise composition of other farms are 26.25 per cent in the category of 101-200 pots, 20 per cent in the category of 201- 300 pots and rest 13.75 per cent in the category of 301 & above pots.

As per the locational features of the sample farms, it is observed that more than 46 per cent farms are located within a distance upto 1 km. from the permanent road and 30 per cent of the farms are located within a distance from 1.01-2kms. The rest of the sample farms are located within a distance from 2.01-3kms. (18.75 per cent) and 3 kms. and above (5.00 per cent).

Out of the eighty growers, more than 21 per cent belongs to the scheduled caste and 10 per cent belongs to the scheduled tribe. It is also fact that the scheduled caste and scheduled tribe growers are mainly marginal farmers. The average size of family of the sample orchid growers varies from 5.4 in large category to 6.4 in marginal category. The average family size in small and medium categories of farms are 6.2 and 5.3 respectively. It is also found that proportion of male worker is more in medium category farms followed by large and small categories and it is lowest in marginal farms.

Similarly, the average work force is found to be more in medium and large farms than that of marginal and small farms. The literacy percentage is also less in marginal and small categories of growers. The overall literacy percentage is found to be 43.87 per cent. The land holding size of the sample farms varies from 2.25 acres in marginal farms to 15.34 acres in large farms. The overall farm size is found to be 5.55 acres. Among the field crops grown by the orchid growers, paddy accounts for the major share (54.99 per cent) and the share of paddy in the cropping patterns found to be highest in medium farms (65.51 per cent). The cropping pattern of these farms other than paddy is concentrated towards cultivation of vegetables. But the share of vegetables in the cropping pattern varies from farm to farm which may be due to difference in irrigation facilities and altitude of land. However, there is no evidence that flowers are grown by the growers other than orchid by the sample farms.

4

EXPORT ORIENTED HORTICULTURAL CROPS

While the Covid-19 pandemic has disrupted the global supply chains in a major way since the beginning of 2020, India has been able to boost exports of agricultural and processed food products despite challenges posed by the pandemic.

The significant growth in production has also led to enhancement of exports of fruits and vegetables to the major traditional markets of UK, United Arab Emirates (UAE), Bangladesh and West Asian countries. Notwithstanding the challenges posed by Covid-19 pandemic, there has been a significant jump in the exports of horticultural crops during 2020-21.

The realisation from India's fruits and vegetables exports rose to Rs 11,019 crore ($1487 million) in 2020-21 from Rs 10,114 crore ($ 1408 million) reported in 2019-20. The value of processed fruits and vegetables exports grew to Rs 10,277 crore ($ million 1386) in 2020-21 from Rs 8315 crore ($ 1164 million) in 2019-20. APEDA has made persistent marketing efforts with European Union (EU) markets resulting in exports growth of more than 9 per cent in 2020-21 over the year 2019-20.

APEDA has been making efforts to boosts exports of fruits, vegetables and plantation crops through taking series of measures including creating adequate infrastructure, ensuring supplies of quality input material, controlling pests and diseases, ensuring compliance of phyto-sanitary norms and establishment of market linkages for boosting exports through collaborations with various institutions.

INFRASTRUCTURE DEVELOPMENT

APEDA has been assisting exporters for setting up modern integrated packhouses and purchase of refrigerated vehicles to support the supply chain in horticulture. This has largely helped exporters of fresh horticulture produce like mangoes, pomegranates, grapes, bananas, fresh vegetables and the floriculture sector. Due to the short shelf life of most fruits and vegetables, more than 50 per cent of exports are through air cargo to destinations such as the UK, Netherlands, Germany, Australia, Japan, Korea, Singapore, Africa and the Middle-East.

The modern integrated pack houses which are able to meet the requirements of issues associated with quality and post-management have helped the horticulture export sector. APEDA has assisted in the establishment of around 250 export oriented European Union norm compliant pack houses in the private sector. The financial assistance provided to the State Government under Infrastructure Development has also enhanced the exports potential.

Clusters based approach

For the implementation of the Agri Export Policy announced in 2018, 14 product clusters spread across in six states dedicated to horticulture products have been identified. The model of Banana exports from the clusters of Anantapur and Theni in Tamil Nadu, Jalgaon and Solapur in Maharashtra have been testimony to this fact. Similar success has been achieved by increasing exports of mangoes from clusters in Andhra Pradesh, Telangana, Uttar Pradesh and Maharashtra. Other successful models have been for grapes from Nashik, vegetables from Uttar Pradesh and Bihar. The farmer centric model of cluster development has taken place in Varanasi, Uttar Pradesh. The export hub project of vegetable and fruit has been able to demonstrate usefulness of backward and forward linkages.

Collaboration with institutions

APEDA is currently collaborating with institutions such as Indian Council for Agricultural Research (ICAR), Council for Scientific and Industrial Research (CSIR) and Central Food Technological Research Institute (CFTRI) for export-oriented research to support Indian Horticulture exports. Linkages with organizations like IFFCO, ITC, Agri Universities & Krishi Vigyan Kendras is also promoting appropriate farm practices.

Market access

The countries do impose the restrictions of market access and sometimes non-tariff barriers to inhibit the imports. The tool is also used to bargain for market access of their produce in India. The horticulture produced in India is subject to competition in international markets from Asian countries like China, Japan, Thailand, Philippines etc. Similarly, competition is also faced from CIS countries and the other leading fresh produce regions from Africa and America. The absence of bilateral agreements between major importers and India has also proved to be disadvantageous. The examples are exports of fruits and vegetables from Africa and Latin America to Europe. These countries enjoy preferential duty concessions and at the same time face no technical barriers while exporting, as compared to India. APEDA proposes aggressive negotiations for bilateral agreement for turning the table in favour of India.

There is a need for better marketing avenues to Indian exporters by undertaking huge branding exercises to place the produce in the shelves of major marketing chains. Efforts have been made by organising product specific and country specific Buyer Seller Meets to help branding of Indian horticulture produce. India has huge potential to further expand the export of horticultural crops. This could boost farmers' income.

ANDHARA PRADESH

The per hectare average maintenance cost of Thompson seedless grapes is Rs. 1,27,795 of which a major proportion is spent on the application of FYM and compost (23.01%), Pesticides (21.15%) and Fertilizer (13.63%). Together, these three items account for nearly 57.79% of the total costs incurred by the sample households. The other important costs incurred are on the maintenance of basins (6.96%), interculture (8.87%), watch and ward (5.88%) and apportioned initial establishment costs of non- bearing orchard (10.00%).

The average per hectare gross and net returns from the Thompson seedless grape reported by the sample households are Rs.3,03,458 and Rs.1,75,563 respectively. However, they varied among different size groups of sample farmers. While the highest gross (Rs.320574) and net (Rs.1,97,956) income are reported by the small farmers followed by gross (Rs.2,95,370) and net (Rs.1,80,958) incomes are reported by the marginal farmers and the lowest gross (Rs.2,92,367) and net (Rs.1,60,754) incomes are reported by the large farmers. This is mainly due to high cost of production incurred by the large farmers and low returns from the crop compared to the small farmers who incurred less costs and secured good harvest.

As already mentioned non-bearing sample have been growing Anab-e-Shahi variety of grapes is not available in the sample villages and information on the initial costs have been collected from the bearing sample households and approtioned among various stages of plantation proportionately. No attempt is made to analyse the costs on non- bearing plantation. Maintenance costs of various stages of bearing and the returns have been analysed in detail for Anab-e-Shahi grapes.

The per hectare average maintenance costs incurred by the sample households of Anab-e-Shahi grapes have come to be around Rs.1,11,460 of which FYM and compost, fertilizer, initial constituted nearly 60.00 per cent while interculture, initial costs and other maintenance operations accounted for the remaining expenditure. The proportions are more or less the same among all the size groups.

While the marginal farmers have spent the lowest amount of Rs. 86.163 on the maintenance of orchards, small farmers have incurred the highest expenditure of Rs.1,17,779 to maintain the orchard.

The highest net income of 2,27,016 is reported by the medium farmers, while the lowest of Rs.1,76,511 is reported by the small farmers, the average net income being Rs.2,05,829 among the total sample households.

Per hectare average maintenance cost of mango is reported to be Rs.8226 of which a high proportion of 28.19 per cent of expenditure is on pesticides followed by fertilizers (19.18%), land preparation (14.70%). Together, these four major items accounted for more than 75.00 per cent of the total maintenance costs incurred by the sample farmers. While the highest per hectare cost of Rs.8940 is reported by the medium farmers, the large farmers reported the lowest cost of Rs.7938.

The average per hectare net income from mango orchards is reported to be Rs.1759 during the reference year due to heavy crop failure. While a highest per hectare net income of Rs.2720 is reported by the medium farmers, a lowest net income of Rs. 1197 is reported by the small farmers. The large and marginal farmers have reported the net incomes ranging in between these two categories.

However, as already seen, considerable variations in the costs and returns have been observed at different stages of fruit bearing plantations.

The per hectare average cost of cultivation of onions is reported to be Rs.9624.75 and the highest per hectare cost of Rs.11,205.54 is reported by the large farmers while the lowest cost of cultivation of Rs.7833.42 is reported by the marginal farmers. As onion is a seasonal field crop, bullock labour constituted the major input accounting for nearly 23.58 per cent of the total costs, followed by fertilizer (21.27%) seed (18.89%) hired human labour (15.06%) and pesticides

(12.75%). The other inputs which shared the remaining costs are FYM (4.76%), tractor charges (2.87%) and irrigation charges (0.87%). Though onion is raised as an irrigated crop during kharif season, because of the presence of monsoon rains which covered most part of the kharif season, the cost incurred on irrigation is reported to be very low.

Among the various size groups of farmers, while the large farmers have spent a higher proportion on hired human labour (25.54%), the marginal, small and medium farmers have spent a higher proportion of their expenditure on bullock labour i.e., 31.32%, 27.34% and 20.08% respectively. Only large farmers have reported using tractor for agricultural operation like ploughing, harvesting and transporting while other farmers have depended on bullock labour. While a gradual increasing trend is observed in the costs on hired human labour as the size of the holding increased, a reverse trend is discernible in the use of bullock labour, fertilizer and pesticides.

Heavy crop damage is reported by the sample households during the reference year because of unseasonal heavy rains at the time of harvesting of onion crop. The entire Kurnool district is reported to have been under the grip of heavy unseasonal rains during the harvesting period leaving the crop fully sub-merged in water causing severe damage. Except the marginal farmers whose lands are situated on elevation, all the other farmers have reported negative incomes from onion crop during the reference year.

The average net income is reported to be Rs.596.37 from onion crop and while a net income of Rs.821.15 is reported by the marginal farmers, the highest negative income of Rs.1240.57 is reported by the large farmers. The medium and small farmers have also reported a negative income of Rs.964.44 and Rs.380.23 respectively from the crop. The agencies involved in assembling and distribution of mangoes and grapes are the growers, pre-harvest contractors, commission agents, wholesale buyers, processing industry, retail traders and consumers.

About 3370 tonnes of grapes, both Anab-e-Shahi and Thompson seedless, have been sold by the sample cultivators in the Rangareddi District to the Commission Agents at farm realising an amount of Rs.390 lakhs during the reference year. Of the total quantity marketed only 7.59 per cent is Anab-e-Shahi variety while the rest of the quantity is Thompson seedless variety of grapes. The average per Kg. price realised by the sample cultivators is Rs.11.59 and it worked at to Rs.10.15 for Anab-e-Shahi and a slightly higher price of Rs.11.71 for Thompson seedless grapes.

In the case of vegetables like onions, the grower himself directly sells the produce to the commission agents either at village point or in the nearest main markets. The commission agents also act as wholesale merchants and often undertake the distribution of produce to distant markets. The commission agents supply the produce to the retail traders who inturn market the produce to the consumers. Sometimes the retail traders also participate in open bidding in the Kurnool assembling centre. The small growers also sell the produce directly to the hawkers, push cart traders etc. in the nearest towns. It is estimated that about 50-60 per cent of the total production of onions is collected and distributed through commission agents, the balance is directly sold to hawkers and small retail traders by the cultivators.

About 194 tonnes of onions have been sold by the sample cultivators to the commission agents in the Kurnool district, of which about 61.16 per cent of the quantity is sold at village point and 38.84 per cent at the market point realising an amount of 3.86 lakhs. The per quintal average price realised by the farmers is around Rs.199 and while it worked out to Rs.198 at village point, it increased to Rs.200 at market point.

BIHAR

The average size of litchi orchard was 1.2 hectare with about 170-200 trees per hectare. The cost of establishment of litchi orchard till its first bearing stage was estimated at Rs.22,276/-. Annual maintenance cost per hectare was estimated at Rs.12,184/-. The results of economic analysis indicate high benefit cost ratio (1.4:1) as well as high internal rate of return 21 per cent and small pay back period again shows less risk in investment in litchi cultivation.

Seasonally in production and consumption of litchi was observed to be its very peculiar character. Its peaking marketing period ranged from first week of May to mid June in year litchi growers leased their standing crop to pre-harvest contractors or village traders for one or more year generally just before flowering and in turn received ready money. Moreover, they lacked expertise essential for litchi marketing and also they wanted to save time, money and labour. Regarding mode of sale, written document were executed between pre-harvest contractors and growers depicting all the terms and conditions including mode of payment before witness. In some cases the written documents were furnished before the Registrar bearing detailed terms and conditions. Pre-harvest contractors and village traders negotiated privately with growers regarding price of the orchard considering number of tree, setting of fruits, maintenance of orchard, age of plant and percentage of bearing fruit etc. In general 25 per cent of total price was paid at the time of making contract and rest 75 per cent was finally paid at the time of harvest.

Marketing system of litchi was observed to be purely a traditional private system comprising litchi merchant viz., pre-harvest contractor, village trader, commission agent and retailer. The pre-harvest contractor was observed to be most important intermediary who performed most of the marketing services like watch and ward of orchards, harvesting of fruit, packing of harvested fruit into boxes, grading and transportation of the consignment right from the orchard to distributing Centre. Litchi was packed into wooden boxes for transportation of distant markets and bamboo baskets were used for packaging of litchi meant for local sale.

No systematic grading was observed to be practised at either pre-harvest contractor or commission agent level except sorting of fruits according to varieties. However, at retailer level fruits were graded into two or three lots mainly depending on size, degree of damage and a differential prices were charged for different lots.

It was observed that litchi produced in the study area was marketed to distant and local markets through three important channels. The identified channels were as follows :-

1. Producer-pre-harvest Contractor-Commission agent (Muzaffarpur)-Wholesaler (Distant market like Delhi)-Retailer-Consumer (Distant market like Delhi).
2. Producer-Village Trader-Wholesaler-Commission age (Muzaffarpur)-Wholesaler (Distant market)-Retailer Consumer (Distant market).
3. Producer-Retailer-Consumer (Muzaffarpur market).

The analysis of price spread in the Muzaffarpur market reveals that pre-harvest contractor's share in consumer's rupee was 62.72 per cent. The purchase price of wholesaler was rupee 687.06 per Qt. The wholesaler paid Rs.137.06 as marketing charges. The highest marketing charges per quintal (Rs.56.49) was recorded for packing of litchi followed by spoilage (Rs.35.33), transportation (Rs.33.22), palledari (Rs.6.67) and octroi (Rs.3.57) per quintal. The miscellaneous expenditure (Rs.1.78) was the lowest on telephones and telegrams, etc.

The retailer's sale price was Rs.1,095.50 per quintal. The charges borne by retailer was Rs.62.28. The total amount that the paid for one quintal of litchi was 945.46. Thus, he earned a net profit of Rs.150.04 per quintal. The total cost of marketing incurred by the contractor/wholesaler worked out to 12.51 per cent of the consumer's rupee which was 68.75 per cent of the total marketing cost. The marketing cost paid by the retailer was 5.68 per cent of the consumer's rupee. The total marketing cost was estimated as 18.20 per cent of the consumer rupee. The producer's share in consumer's rupee was very less due to the fact that producers sell directly their orchards to the pre-harvest contractor/wholesaler. Therefore, more profit is earned by the middlemen. The producer's share in consumer's rupee may increase if the producer can sell their produce (Litchi)bringing it into the market by themselves.

The price spread in distant market (Delhi) explains that the marketing costs incurred by the commission agent was the least being only 1.36 per cent of the consumer's price. The expenditure incurred by commission agent was minimal due to the fact that the pre- harvest contractor bears all the expenditures except market fee, postal and telegraph charges and other petty charges. The marketing cost borne by retailers were also high about Rs.21.7 per box (6.84%) which includes in labour charges, transport, dalali, spoilage and other miscellaneous charges. The total price per box worked out to be Rs.193.32 out of consumer's price of Rs.317.32.

It is also evident from the study that the absolute share of producers in consumer's rupee in Delhi market was Rs.124 per box which was 39.10 per cent of the total consumers price paid. The total marketing costs incurred by pre-harvest contractor, commission agents and retailers in distant market (Delhi) were Rs.77.40, Rs.4.30 and Rs.21.70 per box in total price of Rs. 317.32 per box paid by the consumer. The marketing margin of pre-harvest contractor, commission agents and retailers were Rs.21.47, Rs.15.88 and Rs.52.89 per box. the total marketing cost and margin for all the intermediaries worked out to Rs.193.32.

The study of producers' share in export earning explains that the producers have not been generally involved in export of litchi from the project area. therefore, the selling price is their margin. The exporters' share in export earning is about 11 per cent. However, producers get 5.71 per cent in export earning. So it is worth pointing out that producers are not much benefited by the litchi export directly because the prices they get for their produce are more or less the prices prevail in the local market. Moreover, the export of litchi generates new income and employment opportunity for the whole regions and earns foreign exchange. Therefore, the export of litchi is beneficial for the Indian economy in general and economy of Bihar in particular.

GUJARAT

On the whole for orchards of chiku bearing age group, the per hectare total cost was Rs. 60036 of which production cost was Rs. 52033 (86.7 %) and marketing cost was Rs. 8003 (13.3 %).

The per hectare total cost was maximum at Rs. 65545 with orchards of 26-40 years of age and minimum at Rs. 57665 with orchards of 9-25 years of age.

The production cost was maximum at Rs. 56653 with orchards of 26-40 years age and minimum at Rs. 50090 with orchards of 9-25 years of age.

The marketing cost was maximum at Rs. 8892 with orchards of 26-40 years minimum at Rs. 7575 with orchards of 9-25 year of age.

Overall per hectare gross value of chiku production was Rs. 84162. It was maximum at Rs. 93974 with orchards of 26-40 years of age and minimum at Rs. 79247 with orchards of 9-25 years of age.

On the whole, per hectare net income from chiku cultivation was Rs. 24126. It was maximum at Rs. 28429 with orchards of 26-40 years and minimum at Rs. 21582 with orchards of 9-25 years of age.

With regard to non-bearing orchards, the per hectare net income from intercropping was Rs. 26311, the per hectare expenditure was Rs. 24731, which left a net income of Rs. 1580 per hectare.

Overall per hectare cost of non-bearing orchards was Rs. 24731. Cost of plants, plant supports, rented value of land, irrigation and human labour were the major inputs, constituting over 60 per cent of total expenditure.

Overall per hectare net income from intercropping was Rs. 26311, which had resulted Rs. 1580 as per hectare net income. It was maximum to Rs. 1954 with large farm households and minimum to Rs. 925 with marginal farm households.

On the whole, per hectare total cost of orchards of 9-25 years age of all the sample households was Rs. 57665, of which production cost was 86.9 per cent and marketing cost was 13.1 per cent. Out of total production cost labour (31.7 %), imputed value of owned land (31 %) were the major components of inputs.

The overall per hectare gross return of chiku crop was Rs. 79247. The overall per hectare net income was Rs. 21582. It was maximum at Rs. 23335 with large farm households and minimum at Rs. 14759 with marginal farm households.

Overall, per hectare total cost of the orchards of 26-40 years of age was Rs. 65545 of which production cost was 86.4 per cent and marketing cost was 13.6 per cent. The per hectare total cost ranged from Rs. 68102 with small farm households to Rs. 60981 with large farm households. Labour, imputed valued of owned land and fertilisers were the major components of production cost. On the whole per hectare gross return of the orchards was Rs. 93974. Overall per hectare net income was Rs. 28429. It was maximum at Rs. 29957 with small farm households and minimum at Rs. 26898 with large farm households.

On the whole, per hectare total cost of orchards of 41 years of age and above was Rs. 58732. The production cost was 86.6 per cent and marketing cost was 13.4 per cent of total cost. Overall per hectare gross return of the orchards was Rs. 82862. Overall per hectare net income was Rs. 24130. It ranged from Rs. 34427 with medium farm households to Rs. 15889 with marginal farm households.

Per hectare total cost of the orchards of bearing age groups of all households was Rs. 60036. The production cost was 86.7 per cent and marketing cost was 13.3 per cent of total cost. Labour (32.2 %), imputed value of owned land (31.5 %), farm yard manure and organic manure together (9.8 %) and fertilisers (8.2 %) were the major components of production cost. Overall per hectare gross return was Rs. 84162. It was maximum at Rs. 881143 with small farm households. On the whole, per hectare net income was Rs. 24126. It was at Rs. 26787 with small farm households and at Rs. 22020 with large farm households.

The producer's share in consumer's rupee is estimated on the quantity of chikus sold in Delhi market on March 18, 1998 through Gadat Khedut Vividh Karyakari Sahakari Mandal Ltd.', Gadat, Gujarat.

The quantity of chikus were packed in hard paper boxes of 10 kgs. each and were transported through road transport by trucks. The trucks started from Gadat at 12 noon and reached in Delhi market early in the morning at 4 o'clock, after travelling the journey of nearly 40 hours.

The consumer's price of average quality and size of chikus was estimated to be Rs.130 per box of 10 kgs. The retailer's price was estimated at Rs.110 per box of 10 kgs. The wholesaler's price was Rs.84, while the net price received by an orchardist was Rs.52 per 10 kgs. Thus, the producer's share in consumer's rupee was reported to be 40 per cent.

HARYANA

Floriculture units in the North need to invest substantially in temperature control in the green house. As such, other factors remaining the same, the cost of installing a green house and related equipment is likely to be much higher in North India than that in the other regions. The data presented in suggest that as against an investment of Rs. 658 lakh required for a green house and related equipment in the Northern region, the cost of a green house and related equipment in Southern region is Rs. 335 lakhs. The cost of green house and other equipment as percent of the total project cost, on an average, works out to more than 48 per cent.

The cost of production of a cut flower producing unit would depend on a host of factors. Based on an average of ten floriculture units, the average annual cost of production of a floriculture unit has been estimated to be Rs. 207.54 lakhs at 1995-96 prices. Of this, depreciation and interest cost account for 30 per cent, raw material for 14 percent, wages for another 8 per cent while administrative expenses account for about 20 per cent of the total cost. The remaining costs are accounted for by such items as overheads, freight, royalty etc.

HIMACHAL PRADESH

The per hectare annual maintenance cost of non-bearing orchard is observed to be almost same i.e. more than rupees 14 thousands in Kullu and Mandi districts while the same is more (Rs.18,174) in Shimla district. The major reasons for high cost of maintenance of non-bearing orchards in Shimla district is non performing of inter- cropping and high cost of labour employed. In Kullu district the cost on labour is much higher (Rs.3707/ha.) than in Mandi district (Rs.1730/ha.)and consequently cost of maintenance is also higher in Kullu consequently cost of maintenance is also higher in Kullu district. The returns from inter-cropping are estimated to be Rs. 2753 per hectare in Kullu district. However, after deducting the returns from inter-crops, the annual cost of maintenance of non-bearing orchard in Kullu comes to be the same as in the case of district Mandi where the net returns from inter-crops are meagre (Rs.199/ha.).

The per hectare annual maintenance cost of increasing production stage (8-17 years) of orchard is observed to be high(Rs.53611) in Shimla district and low (Rs.41,727) in Kullu district. Since the productivity of apples is reported to be high in Shimla district as compared to Kullu and Mandi districts, the per hectare cost on marketing is also observed more (Rs.81,718) in Shimla district. The per hectare total cost of production and marketing is estimated to be Rs.1,35,329, Rs.1,13,534 and Rs.1,06,051 in Shimla, Mandi and Kullu districts respectively. The share of marketing cost is estimated to be more than 60 per cent in Shimla and Kullu districts and about 58 per cent in Mandi district. Because of low level of productivity, the net returns from increasing production stage of orchard in Mandi district is observed to be the lowest (Rs.38,807/ha.). The per hectare net returns in Shimla district are observed to be the highest (Rs.54,3211).

The per hectare annual maintenance cost of constant production stage of orchard is estimated to be more (Rs.59,283) in Shimla than in Kullu district (Rs.43,090). Because of better productivity levels achieved in Shimla and Mandi districts the marketing cost incurred is observed more (above 103 thousands) than in Kullu district (Rs.69,459/ha.) With regard to cost of production and marketing, the Mandi district is leading in net returns (Rs.87,791/ha.)from constant production stage of apple orchards. Though the returns in Shimla district are more (Rs.88,372/ha.) but the input-output ratio is lower than in Mandi. The reasons for this trend is due to early attainment of full bearing in Mandi district. In Kullu district the net returns from above stage of orchards are estimated to be Rs.70,203 per hectare.

The per hectare annual maintenance cost of decreasing production stage (above 30 years) of orchard is estimated to be Rs.43,860, Rs.50,210 and Rs.59,580 in Kullu, Mandi and Shimla district respectively. Because of high elevation and relatively better climatic conditions the productivity achieved in Shimla district is far better than Kullu and Mandi districts, and consequently the net returns are observed to be more than 157 per cent of Mandi district. In Kullu district the net returns from above category of orchards are observed to be 40 per cent less than Shimla district which are Rs.1,30,100 per hectare.

On the whole the foregoing analysis suggests that the annual maintenance cost of non- bearing and bearing orchards is observed to be more in Shimla district and low in Kullu district. Similarly per hectare marketing cost is also observed more in Shimla district and low in Kullu district. Similar trend is also observed in net returns from apple cultivation in the State. However, the input-output ratio shows that per rupee returns from all age of orchards are more in Kullu district than in Mandi and Shimla districts.

Upto nineties the phenomenon of selling the standing crop to pre-harvest Contractors was common in fruit growing areas of Himachal Pradesh. Now 5 and 7 per cent of the sample orchardists in Mandi and Kullu areas sell their crop to pre-harvest contractors.

While in Shimla district producers themselves or through Co-operative Marketing Societies undertake the marketing of their produce, at the state level the ratio is 12 per cent only. The pre-harvest contractors undertake the entire marketing functions such as picking of fruits, grading, packing, arrangement of transportation and selling of produce etc.

In apple, important components of marketing cost are labour used in picking, assembling, grading, packing and packing material, transport and market charges and taxes etc. In the present study Delhi Azadpur market has been taken as the representative market to work out the producer's share in consumer's rupee. During 195-96 the arrival of apples from Himachal pradesh in Delhi Azadpur market was to the tune of 209735 tonnes which accounted for 30.90 per cent of total arrival of apples in Azadpur market. The above quantity of apples is 80.77

per cent of total apple sold outside the state during 1995-96. Out of the total marketing cost involved in apple crop, the preparation of fruits accounts for 45 per cent, transportation for 32 per cent and share of market charges and taxes is estimated to be 23 per cent.

The poor quality and culled type of apples are also sold in gunny bags. This practice has been observed in Shimla district only. Generally apple packed in gunny bags are sold locally and to the traders visiting the producing area.

The analysis of marketing cost of apples suggests that per box (about 18 kgs) cost is estimated to be Rs.78.4 at the State level and the same is observed more (83.7) in Shimla district. The per kg. cost of marketing of apples is estimated to be more than Rs.4.1 in all the selected districts and across the farm sizes.

At the State level the producers share in consumer rupee is estimated to be 47.7 per cent and the same is observed more (48.0 per cent) in Kullu district. Across the farm size, medium farmers are realising larger share (47.0 per cent) than other categories of farmers. The per box (about 18 kgs.) consumer price of average quality and size of apples is reported to be Rs.241 during 1996 marketing season (Azadpur Market, Delhi).

KERALA

Cashew is a perennial crop with a wide span of 40 years life period. The cost of cultivation and the income are calculated keeping into account the whole life of the tree.

The main items of cost are manure, fertilisers, plant protection measures, wages, land revenues, repairs to implements and interest on loan etc. On an average Rs.8040/- per hectare has been spent towards cost of cultivation during the reference period. In this nearly 50 per cent of the expenditure accounts to wages and harvesting. Among the groups the small farmers have incurred comparatively less cost than the medium and big farmers.

The average yield per hectare comes to 952 kg. The small farmers have obtained the highest yield of 987 kgs., followed by the medium farmers 905 kgs. and big farmers 902 kgs.

The average gross income per hectare accounts to Rs.33,325/- and the net income works out to Rs.25,285/-. The net income is calculated at the excess of gross income over the cost of cultivation. The small farmers raised their income more than the medium and big farmers because of their relative advantage of family labour and personal participation and involvement in farming activities.

Private millers are the sole purchasers of raw nuts. All the farmers disposed of their produce to the millers at the village itself. The farmers have not incurred any transport cost to market their produce so the selling price remained constant and no price variations prevailed in the villages.

The average cost of cultivation for all the 40 years for a hectare is estimated to Rs.2,90,516. Which includes gestation period, pre-full and full development periods and accounts per hectare an average cost of Rs.7,263/- per year. The average amount spent for the gestation period per hectare works out to Rs.5,181/- per year and Rs.4,286 for pre-full development period. The medium and big farmers have incurred higher cost because of the payment of wages for hired labourers.

MAHARASTRA

A comparative analysis of cost structure revealed the total cost of onion production to be much higher in rabi season compared to that during kharif season. This held true for all the categories of onion producers. The per acre total cost of onion production during kharif season was estimated at Rs. 12,667 for marginal category, Rs. 11,850 for small category, Rs. 12,532 for medium category and Rs. 11,313 for the large category. As for rabi season, the estimated per acre total cost of onion production turned out to be Rs. 13,674 for marginal category, Rs. 12,551 for small category, Rs. 12,827 for medium category and Rs. 13,352 for the large category. In general, the per acre total cost of onion production for an average category of farmer in the kharif season was estimated at Rs. 11,678 and the same figure during rabi season stood at Rs. 12,949. During the kharif season, the production cost as proportion of total cost varied from 80 per cent in the case of medium category to 85 per cent for the small category. In rabi season too, no significant difference in production cost as proportion of total cost was observed for various categories of onion producers with this proportion varying from 79 per cent for large category to 85 per cent for the medium category. The per acre production cost was seen to decline with the increase in size of land holding of the sample producers, though the difference in per acre cost of production for the small and the medium categories was not very high. Both the seasons had preponderance of this trend. As for the per acre marketing cost, it was highest for the medium category in kharif season and for large category in rabi season; lowest per acre marketing costs were recorded by the small category producers in both the seasons.

The labour cost as proportion of the total cost was observed to decline steadily with the rise in land holding size of onion producers, though the decline in this proportion being not very sharp. As against this pattern, the material input cost as proportion of the total cost was found to increase with increase in the land holding size of the producers in the kharif season; the increase being from 14 per cent in the case of marginal category to nearly 18 per cent in the case of large category. In the rabi season, the material input cost as proportion of the total cost turned out to be the highest for medium category (23 per cent), followed by large and small category (about 20 per cent) and then by marginal category (about 18 per cent) in that order. Most of the material input cost owed it to investments made on the purchase of fertilisers. While the large category of onion producers invested nearly two and a half times more on the fertiliser input in the kharif season compared to the marginal category, this investment in the rabi season was observed to be 75 per cent higher for the medium category of onion producers as compared to those belonging to the marginal category.

In general, it was observed that the average category of onion producer spent nearly 70 per cent of the gross operational cost of onion production on cultivation related activities such as ploughing (6 per cent), sowing (10 per cent), manuring (8 per cent), fertilising (16 per cent), interculture operations (8 per cent), mulching (1 per cent), irrigation and maintenance (10 per cent), plant protection measures (3 per cent) and harvesting operations (8 per cent), various marketing functions accounted for the rest of the 30 per cent investment of the gross operational cost of onion production and this included expenditure on family as well as hired labourers engaged in various marketing related operations.

During the kharif season, the per acre gross return from onion cultivation was estimated at Rs. 16,829 for marginal category, Rs. 17,332 for small category, Rs. 17,613 for medium category and Rs. 16,272 for the large category. During the same season, the per acre net return was estimated at Rs. 4,162 for marginal category, Rs.,842 for small category, Rs.5,081 for medium category and Rs.4,959 for the large category of onion producer. the higher per acre net return witnessed in the case of small category during the kharif season was mainly due to lower per acre total cost of production for this category compared to other categories of producers. The gross per acre returns in the rabi season for various categories of onion producer was estimated at Rs.18,574 for the marginal category, Rs. 18,589 for small category, Rs. 18,562 for medium category and Rs. 20,442 for the large category. and, the net per acre return in the same season was estimated at Rs. 4,901 for the marginal category, Rs. 6,038 for small category, Rs. 5,735 for medium category and Rs. 7,090 for the large category. Thus, one acre onion farm yielded much higher net returns for the small category of producers during the kharif season and, for the large category of producers in the rabi season. In general, for an average category of onion producer, the per acre net return was estimated at Rs. 5,098 in the kharif season and at Rs. 6,282 in the rabi season. Thus, the average category of producer was seen to obtain 23 per cent higher net return from one acre onion farm during the rabi season as compared to kharif season. In this sequel, the input-output ratio was seen to be higher in rabi season (1:1.49) compared to kharif season (1:1.44).

The total per household production of onion for an average category of producer during the kharif season was estimated at 584.19 quintals, of which 85.36 per cent or 502.92 quintals was sold as marketed surplus. Although the percentage sale of production remained constant for various categories of producers, a steady decline in home consumption as proportion of production could be observed with the increase in land holding size of onion producers. Nonetheless, in absolute terms, the quantity used for home consumption was seen to be much higher for large category compared to marginal and other categories of onion producers. During the rabi season, the per household production of onion for an average category of producer was estimated at 631.58 quintals, of which nearly 85 per cent, that is, 538.13 quintals was sold in the market as surplus.

In the case of all-bearing age grape orchards, the per acre annual gross maintenance cost was estimated at Rs. 53,828 for marginal category, Rs. 59,431 for small category, Rs. 66,473 for medium category and Rs. 63,795 for large category of orchardists. This indicated that the medium category of orchardists spent 23 per cent more than those in the marginal category on the maintenance of every acre of grape orchard. Upon splitting this gross maintenance cost of production into various component costs, it was noticed that among various categories of orchardists, about 67 per cent of the gross maintenance cost was spent on various production related operations and the remaining 33 per cent owed it to investments are various marketing functions. Among various components of production cost, the prorated establishment cost alone accounted for about 34-35 per cent of the gross maintenance cost of production. The share of material input cost in gross maintenance cost of production was seen to be of the order of about 10-11 per cent. This held true for all the categories of orchardists. Similarly, the share of labour cost in gross maintenance cost for all categories of orchardists was found to be about 6 per cent. As regards the marketing cost, packing material alone accounted for 10 per cent of the gross maintenance cost. Nearly 5-6 per cent of the gross cost of grape production was found to have been spent towards labour cost hired to facilitate various marketing functions. An increase in expenditure on various items of costs, both production and marketing, was seen with the increase in land holding size of orchardists.

The per acre gross annual return for various categories of grape orchardists was estimated at Rs. 76,934 for marginal category, Rs. 92,755 for small category, Rs.96,442 for medium category and Rs.95,968 for the large category. On the other hand, the per acre net annual income from grape orchard was estimated at Rs. 23,107 for marginal category, Rs. 33,324 for small category, Rs. 29,969 for medium category and Rs. 32,173 for the large category of grape producer. Thus, while the small category grape orchardist was observed to garner maximum returns in the form of net annual income accruing from per acre of grape farm, the return for the marginal orchardist was the lowest. In fact, per acre earnings of small category of grape orchardist were found to be 44 per cent more than the marginal orchardist.

The per acre annual gross maintenance cost of grape orchardists increased sharply during the phase the production rose before levelling off to a constant stage and, thereafter, it declined. the per acre gross and net annual returns for the average category of orchardists also found to follow a trend similar to that of per acre gross maintenance cost, that is, an increase in gross and net annual returns was observed from the increasing to constant production stage which declined rather sharply during the declining production stage. The prorated establishment cost of the orchardist increased in every successive stage of production even though the per acre production cost showed declines from the constant to declining production stage after showing an increase in the same from increasing to constant production stage. The share of labour cost and material input cost in gross maintenance cost of production declined from increasing to constant and constant to decreasing production stage, though in absolute terms the expenditure on labour employment and on various material inputs had increased from increasing to constant production stage. A similar trend was noticed in the case of expenditure of the orchardist on packing material. The share of packing material cost in gross maintenance cost of production declined with every successive stage of production despite the fact that in absolute terms the expenditure of the orchardist rose sharply on packing material from increasing to constant production stage, though, thereafter, a steep decline in the same was also seen. In general, during each stage of production, the expenditure on various items of cost, both production and marketing, increased with the increase in land holding size of the orchardists, that is, big orchardists invested more on various items of cost compared to small orchardists.

The average per orchardist production of grapes was estimated at 270.47 quintals, of which on-farm spoilage (post-harvest loss) accounted for 11.77 per cent, 1.03 per cent was used up in home consumption, 2.12 per cent was bartered away as gratis to labourers working on the farms as payment in kind and remaining 85.08 per cent was sold in the market. As the land holding size of orchardists increased, it was interesting to note that the share of spoilage in the total production registered declines. Similarly, the payment made in kind to labourers and the share of home consumption as proportion of total production also declined with the rise in land holding size of the orchardists.

The practice of getting into pre-harvest contract for their orchards was seen to be very common among orchardists in the study area. But of 50 sample grape orchardists, as many as 21 had given their orchards on contract. At least five reason were cited by the sample orchardists for giving their orchards on contract. The most prominent among these reasons were: avoiding labour problem during marketing operations, avoiding problems associated with marketing and avoiding risks and uncertainties associated with harvesting, pricing and other marketing functions. Contractors were found to lay down their own terms and conditions before contracting orchardists. In general, the contract of the orchard was taken on the basis of quality of the orchard, and the variety of grapes grown in the orchard was considered as the main criterion in finalising the value of the contract.

The per box (4 kgs.) total marketing cost was estimated to be the highest when the produce was sold through forwarding agents in the wholesale market compared to produce sold through other marketing channels. Interestingly, orchardists secured least per box marketing cost when they sold the product directly to wholesalers. In the case of grapes marketed cost was estimated at Rs.12.05 for the marginal category, Rs.10.76 for small category, Rs.12.79 medium category and Rs.12.58 for the large category with an overall average of Rs. 12.35 for the average category of orchardist. Of the total cost of marketing, nearly 40 per cent was accounted for by the preparation of fruits alone. All categories of orchardists spent about 6 per cent of their total marketing cost on local transportation. The share of various marketing charges put together in the total marketing cost also did not vary much among various categories of orchardists and this share hovered around 55 per cent.

All the sample onion producers were also found to employ forwarding agents for trading their produce in the wholesale market. In kharif season, the per quintal cost of marketing of onion varied from Rs.19.92 in the case of small category to

Rs.29.49 in the case of medium category. As for marginal and large category of producers, no significant difference in total cost of marketing of onion was seen and, in these two cases, the estimated total cost of marketing revolved around Rs.25 per quintal. Transportation, commission and such other marketing charges accounted for bulk of the total marketing cost. In general, various marketing charges accounted for 73 per cent of the total cost of marketing in the kharif season. In rabi season too, the expenditure on preparation of vegetable accounted for 27 per cent of the total cost of marketing and the remaining 78 per cent owed it to various marketing charges.

The producer's share in consumer rupee was found to be very high for all the categories of grape orchardists. This share was seen to vary from 55.71 per cent for the marginal category to 60.46 per cent for the medium category. The average category of orchardists had an overall average of 59.49 per cent share in the consumer rupee. Although all categories of grape orchardists had a fairly high share in the consumer rupee, such was not the case in the case of onion producers. In general, for the average category of onion producers, the producers' share in consumer rupee was estimated at 44 per cent in the case of rabi season and it stood at 47 per cent during kharif season.

SIKKIM

On the ground of examining the economics of production and input use pattern of orchid and other field crops, the concept of gross income, net income over different cost structure of these crops are applied. In this connection, the variable cost plus fixed cost (V+F) cost B, cost C alongwith net income over this cost i.e. in NIVF, NIB and NIC are worked out. In addition to these the output-input ratio over these costs are also worked out. It is found that cost C of orchid per pot varies between Rs.343.74 in small farms to Rs.345.18 in medium farms. The overall cost C of orchid is estimated to be Rs.341.71 per pot. The overall gross return over cost C from orchid is estimated to be Rs.208.29 per pot and the overall output-input ratio over cost C is found to be 1.61. the overall cost C and output input ratio over cost C of paddy, maize and wheat are Rs.6107.14, Rs.6997.14 and Rs.1852.17 per acre, and 1.19, 1.23 and 1.27 respectively. In case of vegetables like potato, cabbage, cauliflower and cardamom, the overall output-input ratio is found to be 1.23, 1.63, 1.40 and 1.28 respectively.

In terms of marketing of orchid and other field crops, it is found that the orchid growers are habituated in selling most of the agricultural produce within the village itself except potato. It is reported that in case of potato 100 per cent of the produce is sold in same village. It is also reported that 100 per cent of the produced cabbage, cauliflower and cardamom is sold in small town because most of the wholesale distributors are located at towns. In case of orchid, it is entirely sold to the nearby nurseries to whom they are attached. Taking all vegetables together, the channel wise distribution of growers show that the channel - II(b) comprising of Producers - Commission agents - Aratdars - Processors - Retailers - Consumers is the best channel according to the farmers' preference. But according to the price spread over different channels, the channel - II (c) comprising of producers - Aratdars - Processors - Consumers has proved to be efficient in comparison to the other channels.

TRIPURA

Precise record of cost of cultivation involved in pineapple cultivation are not available for the state. The field investigation showed that cost involved in initial plantation are on labour for jungle clearing, digging of holes for planting and the cost of planting material. Per hectare initial costs are worked out at Rs.7,702/- only. So far as total costs of production is pineapple cultivation is concerned 73.76 per cent of costs involved in labour and remaining 26.24 per cent on planting materials.

Pineapple is a highly perishable commodity and it requires immediate sale after harvest. It involves a number of marketing functions like assembling, standardisation or grading, packaging, transportation, storing and dispersion. In the process of marketing, a number of market functionaries like village traders/ commission agents, wholesalers, retailers, co-operative agencies etc. are involved who traded in different marketing channels.

The marketing channels varied from place to place and crop to crop. The identified marketing channels for pineapple in the study area are :

1. Grower - Commission Agent/Village level trader - Wholesaler - Retailer - Consumer.
2. Grower - Wholesaler - Retailer - Consumer.
3. Grower - Retailer - Consumer.

The truth of the situation is that about 99 per cent of pineapple produced are marketed of which a major part (74.64%) is marketed through the traders and the rest are procured by NERAMAC (25.36%).

The analysis of price spread revealed that there is a wide gap between the price received by the producer and the price paid by the consumer. In Channel - I the share of growers is just 41.97 per cent, in channel - 2 growers net share has been found at 40.50 per cent and in Channel - 3 it is 46.16 per cent. The analysis of price spread revealed that wholesalers margin is higher (17.22%) in channel - 2 than channel - 1 (8.65%).

The commission agents margin in channel - 1 is worked out at 16.05 per cent. The retailers net margin is highest in channel -3 (28.20%) followed by channel - 2 (13.92%) and in channel - 1 (12.34%).

UTTAR PRADESH

Regarding economics of mango production, the initial cost of plantation, particularly material inputs was comparatively higher than the cost of labourers. Among the material inputs the cost of seedlings/saplings was higher than all other material inputs. Among the maintenance costs the maximum costs have been invested on preparation and maintenance of basin, interculture and gap filling.

The output from various crops including mango crop, it has been found that the output from mango has been much higher than all other crops. The utilization of production of food crops as well as the value of production utilised was comparatively on the mango farms of decreasing production stage. While in case of mangoes the maximum amount of production has been found to be marketed on the farms of all the categories. The value of output per plant has been found decreasing with the age of plants.

The total mango production was found to be marketed by the contractors only. The income per farm from the other sources was found increasing with the increase in the stages of mango production. Thus, old mango orchards were found to be more profitable.

Regarding marketing of mango the utilization of hired labourers was higher on the farms of mango growers. Picking and assembling were the marketing operations on which the maximum utilization of human labour was done. The utilization of labour was found increasing with the age of mango plants.

The analysis on marketing costs involved in case of mangoes shows that the mangoes of constant production bearing stage have been found to be costlier than the mangoes of the other two bearing stage. Among the marketing costs involved the other costs have been found to be the highest and in that the freight charges have been found to be the highest i.e. Rs.63.60 per quintal while the average total marketing cost per quintal has been estimated to Rs.124.28.

The price spread of mango shows that the margin of contractor including price paid to owner has been accounted to Rs.1094 per quintal. While the price paid by the consumers to the retailer has been accounted to Rs.1500 per quintal. Thus, the contractors share including the share of the owner of the orchards in the consumers rupee has been estimated to 86.67 per cent.

As regards the economics of potato production it has been found that the expenses have been higher on marginal and large farms in comparison of small and medium farms of potato growers on an average. Potato has been found to be comparatively more profitable than other crops grown on the farms of potato growers. The net income has been estimated to be the highest in case of medium farmers followed by marginal farmers.

The output from crops of different seasons indicates that other crops (fodder etc.) paddy, and urd have been found to be more productive on the farms of potato growers. On an overall rabi seasons crops have significantly given the maximum output and that too because of potato crops. Sunflower along with Zaid vegetables have been found to be the important and emerging Zaid crops in this area. Along with the potato cultivation, business as well as other sources have been found to be prominent sources of income in the area under study.

The analysis on marketing of potato indicates that on an overall average 254.66 quintals of potato per farm has been found to be sold through all the channels. Two marketing channels were commonly adopted in the selected district for the sale of potato. The total marketing cost on an overall average has been estimated to Rs.25.61 per quintal and it was decreasing with the increase in the size of farms. Among the various marketing costs, the cost of bags and grading charges have been found to be the main marketing costs.

The price spread of potato at the levels of various market functionaries indicates that the producers margin including cost of cultivation has been estimated to Rs.189.25 per quintal. The price paid by retailers to wholesalers have been reported to Rs.280.00 per quintal. The price paid by consumers to the retailers have been reported to Rs.300.00 per quintal. Thus, the producers share in consumers rupee has been estimated to 86.87 per cent on an average.

WEST-BENGAL

On the ground of examining the economics of production and input use pattern of orchid and other field crops, the concept of gross income, net income over different cost structure of these crops are computed. In this connection, the variable cost plus fixed cost (V+F) cost B, cost C alongwith net income over this cost i.e. NIVF, NIB and NIC are worked out. In addition to these the output-input ratio over these costs are also worked out. It is found that cost C of orchid per pot varies between Rs.329.32 in medium farms to Rs.327.27 in marginal farms. The overall cost C of orchid is estimated to be Rs.317.66 per pot. The overall gross return over cost C from orchid estimated to be Rs.203.39 per pot and the overall output-input ratio over cost C is found to be 1.64. The overall cost C of paddy is found to be Rs.6970.76 and the overall output-input ratio is 1.26. In case of vegetables like potato, brinjal, lady's finger, cabbage, cauliflower and tomato, the overall output-input ratio is found to be 1.23, 2.21, 1.52, 1.63, 1.40 and 1.57 respectively.

In terms of marketing of orchid and other field crops, it is found that the orchid growers are habituated in selling most of the agricultural produce within the village itself except potato. It is reported that 100 per cent of the produced potato is sold in small towns because most of the wholesale distributors are located at small towns. In case of orchid, it is entirely sold to the near by nurseries to whom they attached. Taking all vegetables together, the channel wise distribution of growers show that the channel comprising of Producers - Wholesalers - Consumers is the best channel according to the farmers' preference. But according to the price spread over different channels, the channel comprising of producers - Faria - aratdars - Retailers - Consumers has proved to be efficient in comparison to the other channels.

5

GROWTH IN EXPORT OF AGRICULTURAL PRODUCTS FROM INDIA

The agriculture sector is the largest source of livelihood in India. The country is one of the largest producers of agriculture and food products in the world. In 2021-22, India's agriculture sector growth rate was estimated to be at 3.9% as compared to the 3.6% in the previous year. The country produces many crops and food grains such as rice, wheat, pulses, oilseeds, coffee, jute, sugarcane, tea, tobacco, groundnuts, dairy products, fruits, etc.

India's foodgrains and sugarcane production (million tonnes)

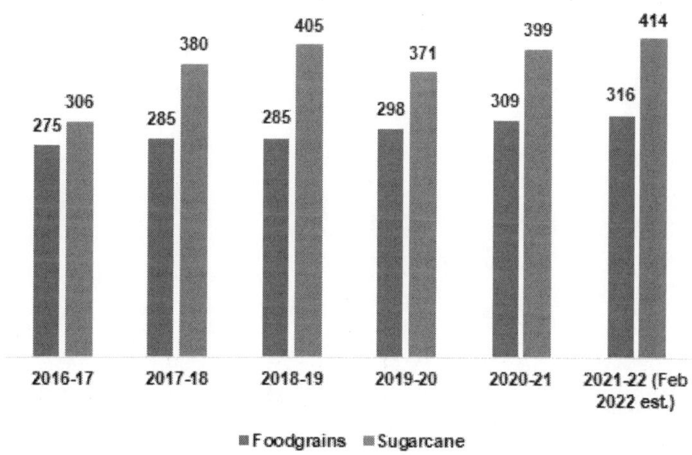

Source: RBI Handbook of Statistics, Ministry of Agriculture and Farmers Welfare

During 2020-21, India's tea production stood at 1,280 million kg. Coffee production during the same period was 354 million kg, a 19% YoY increase. During 2021-22, oilseeds production of India crossed the estimated 37.15 million tonnes while other products such as rice, wheat, maize, pulses, mustard, and sugarcane reached a record high production.

The top crops producing states in India are West Bengal, Uttar Pradesh, Punjab, Gujarat, Haryana, Madhya Pradesh, Assam, Andhra Pradesh, Karnataka and Chhattisgarh. Most of the wheat produced in the country comes from Uttar Pradesh, Punjab, Haryana, Madhya Pradesh, Rajasthan, Bihar and Gujarat. Uttar Pradesh is the largest producer of sugarcane in India contributing about 48%, followed by Maharashtra and Karnataka at 23% and 9% of the total production respectively.

India's agriculture exports trend (US$ billion)

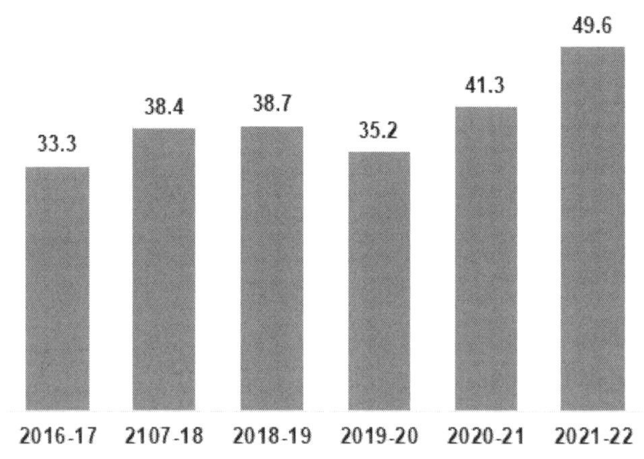

Source: The Ministry of Commerce & Industry

India is one of the largest agricultural product exporters in the world. During 2021-22, the country recorded US$ 49.6 billion in total agriculture exports with a 20% increase from US$ 41.3 billion in 2020-21. India's agriculture sector primarily exports agri & allied products, marine products, plantation, and textile & allied products. Agri & allied products exports were valued at US$ 37.3 billion, recording a growth of 17% over 2020-21.

Rice is the largest exported agricultural product from India and contributed to more than 19% of the total agriculture export during the year 2021-22. Sugar, spices and buffalo meat are among the largest exported products with the contribution of 9%, 8% and 7% to 2021-22 agriculture exports respectively. Wheat exports were valued at US$ 2.1 billion in 2021-22 after recording significant growth over exports of US$ 568 million in 2021-22. Coffee exports have crossed US$ 1 billion for the first time, which has improved realizations for coffee growers in Karnataka, Kerala and Tamil Nadu. Higher exports of marine products, at US$ 7.7 billion, is benefitting farmers in the coastal states of West Bengal, Andhra Pradesh, Odisha, Tamil Nadu, Kerala, Maharashtra and Gujarat.

The government's commitment to increase farmers income can be seen through significant rise witnessed in Agri-exports through giving thrust on boosting exports. Various initiatives taken by the government through APEDA such as organizing B2B exhibitions in different countries, exploring new potential markets through product specific and general marketing campaigns have worked as catalyst to growth exports. The government of India has created products matrix for 50 agricultural products with strong export potential and recognized 220 labs to provide services of testing a wide range of products to enable exporters across India.

The largest importers of India's agricultural products are USA, Bangladesh, China, UAE, Indonesia, Vietnam, Saudi Arabia, Iran, Nepal and Malaysia. The other importing countries are Korea, Japan, Italy, and the UK. During 2021-22, USA was the largest importer of Indian agricultural products at US$ 5.7 billion with share at 11.5% of the total exports. Bangladesh is the major importer of Agri & allied products at US$ 3.8 billion followed by UAE at US$ 2.3 billion. USA and China are major importer of India's marine products.

Country-wise approximate share of exports (2021-22)

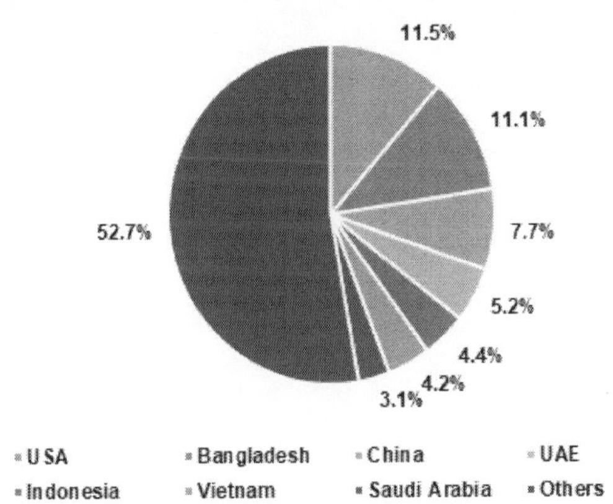

Source: The Ministry of Commerce & Industry

The government of India is keen to organize virtual buyer seller meets (V-BSM) on agricultural and food products with the major importing countries across the world to promote geographical indications (GI) registered with agricultural and processed food products in India. So far 17 V-BSMs have been organized with Kuwait, Indonesia, Switzerland, Belgium and Iran. Similar programs have been organized for Canada (Organic products), UAE & USA (GI products), Germany, South Africa, Australia, Thailand, Oman, Bhutan, Azerbaijan and Qatar.

The government has set up thirteen Agri-Cells in Indian embassies in Vietnam, USA, Bangladesh, Nepal, UAE, Iran, Saudi Arabia, Malaysia, Indonesia, Singapore, China, Japan and Argentina to provide inputs on real time basis to improve Indian exports at these destinations by promoting trade, tourism, technology and investment goals.

According to market research the global food market is expected to bring in revenue of $3.03 trillion by 2020, registering a compound annual growth rate of 4.5 per cent from 2015 to 2020 (APEDA, 2014). It is no surprise that processed foods are an integral part of 10 essential sectors of Make in India initiative.

India as a country is endowed with fertile land, good irrigation facilities and modern technology in the field of cultivation and production of Agriculture/ Horticulture based produce. This advantage is further augmented with development of food processing industry in the country. All these positives related to agriculture/horticulture sector make export of agro products a lucrative proposition for farmers and exporters.

India exported agricultural products to various countries worth Rs 2.10 lakh crore during the year 2015-16. The major agricultural products and their contributions in total agricultural export are provided in Table 5.1.

Table 2.1 Export of major agricultural products from India (2015-16)

Rank	Product Name	Rs Crore	% to Total	Rank	Product Name	Rs Crore	% to Total
1	Marine Products	31183	14.83	20	Sesame Seeds	3012	1.43
2	Buffalo Meat	26682	12.69	21	Tobacco Manufac.	2076	0.99
3	Basmati Rice	22714	10.80	22	Alcoholic Beverages	2027	0.96

Rank	Product Name	Rs Crore	% to Total	Rank	Product Name	Rs Crore	% to Total
4	Spices	16374	7.79	23	Processed Vegetables	1696	0.81
5	Non-Basmati Rice	15086	7.17	24	Dairy Products	1682	0.80
6	Cotton Raw	12816	6.09	25	Pulses	1655	0.79
7	Sugar	9772	4.65	26	Other Cereals	1589	0.76
8	Coffee	5123	2.44	27	Cocoa Products	1260	0.60
9	Cashew	5025	2.39	28	Milled Products	1075	0.51
10	Fresh Vegetables	4763	2.26	29	Wheat	972	0.46
11	Tea	4719	2.24	30	Other Oil Seeds	963	0.46
12	Castor Oil	4616	2.20	31	Sheep/Goat Meat	838	0.40
13	Tobacco Un-manu.	4371	2.08	32	Poultry Products	767	0.36
14	Groundnut	4039	1.92	33	Vegetable Oils	518	0.25
15	Fresh Fruits	3918	1.86	34	Mollases	513	0.24
16	Processed Fruits	3761	1.79	35	Fruits / Vege. Seeds	485	0.23
17	Guargum	3613	1.72	36	Floriculture	478	0.23
18	Oil Meals	3484	1.66	-	Other	3291	1.57
19	Cereal Preparations	3312	1.58		Total	210273	100.00

Source: APEDA

Amongst all agro products, marine products ranked first with export of Rs 31183 crore contributing 14.83 % in total agricultural export of India. It was followed by buffalo meat (12.69%), basmati rice (10.80%), spices (7.79%), non-basmati rice (7.17%), cotton raw (6.09%) and so on.

Table 5.2 CAGR of exports of major agro-products (2009-10 to 2015-16)

Sr No.	Products	CAGR %
1	Marine Products	21.91**
2	Spices	15.88**
3	Cotton Raw (incld. waste)	2.41 NS
4	Fresh Vegetables	11.00*
5	Castor Oil	11.14*
6	Groundnut	14.83 NS
7	Fresh Fruits	18.40**
8	Processed Fruits & Juices	13.42**
9	Sesame Seeds	13.71*
10	Processed Vegetables	15.57**
11	Dairy Products	18.95 NS
12	Poultry Products	13.70**
13	Floriculture	9.14**
	India's Total Agri-Exports	16.48*

** Significant at 5% level* *** Significant at 1% level*

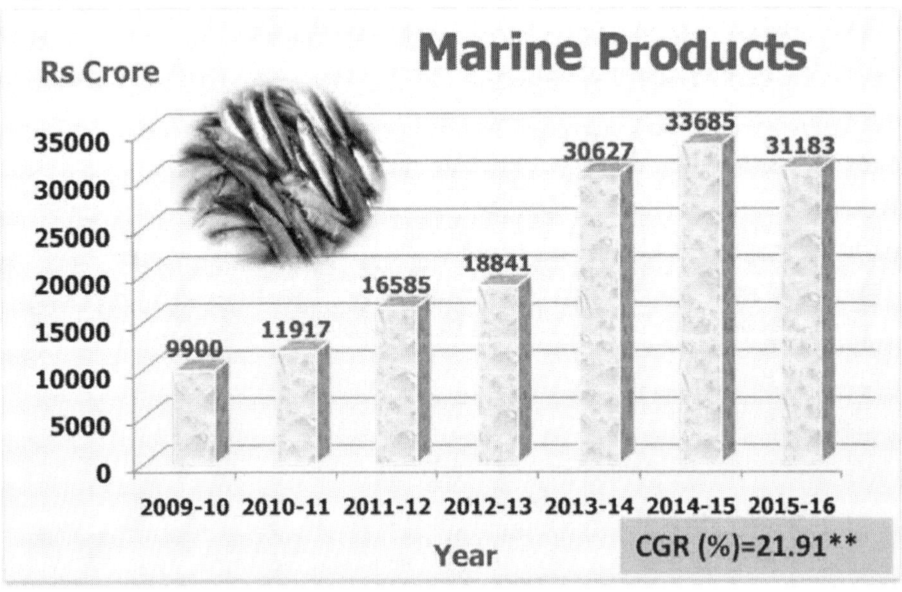

Fig. 5.1 Export of marine products from India

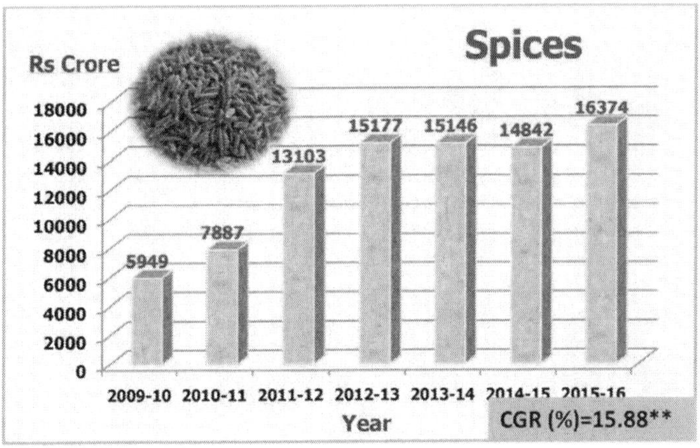

Fig. 5.2 Export of spices from India

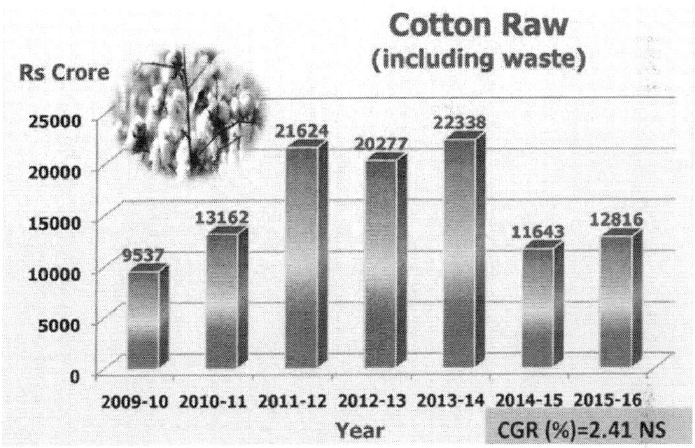

Fig. 5.3 Export of cotton from India

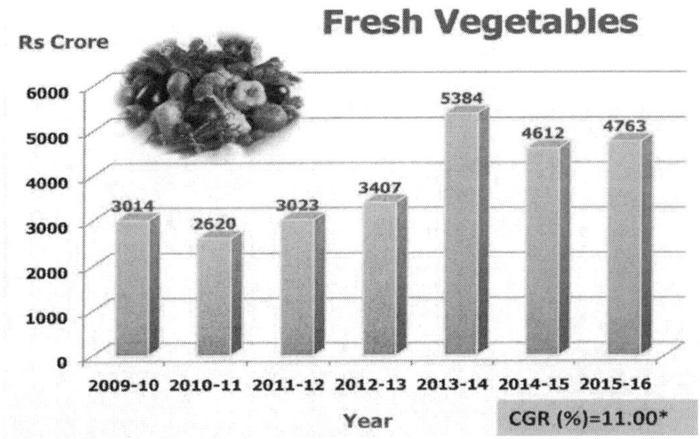

Fig. 5.4 Export of fresh vegetables from India

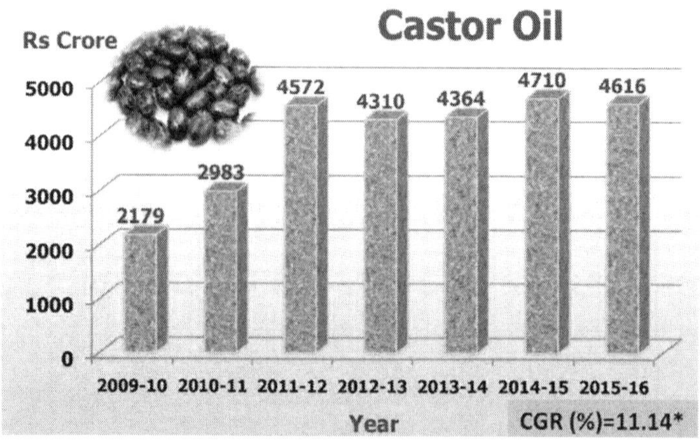

Fig. 5.5 Export of castor oil from India

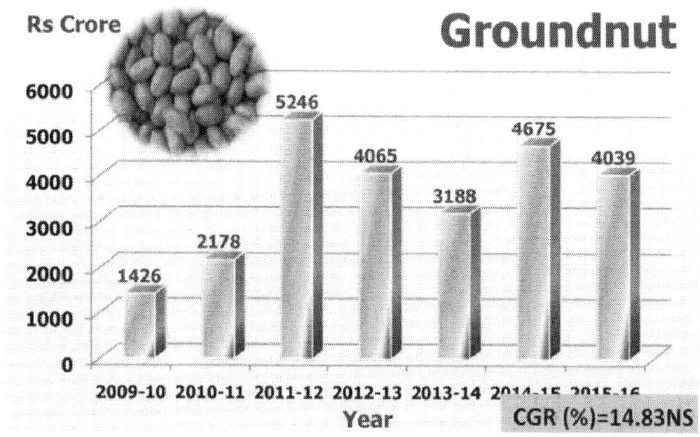

Fig. 5.6 Export of groundnut from India

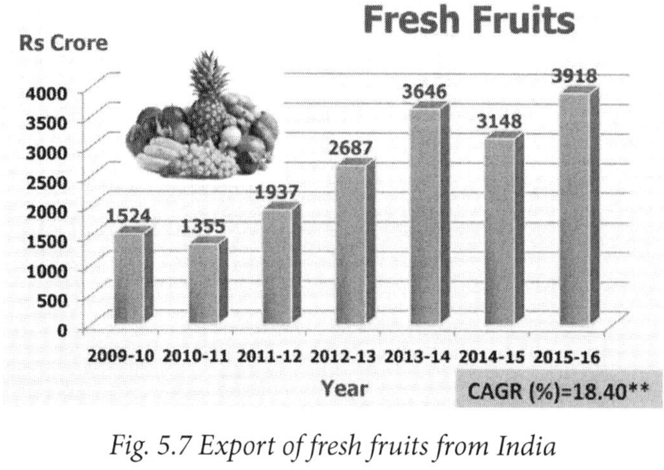

Fig. 5.7 Export of fresh fruits from India

Fig. 5.8 Export of processed fruits and juices from India

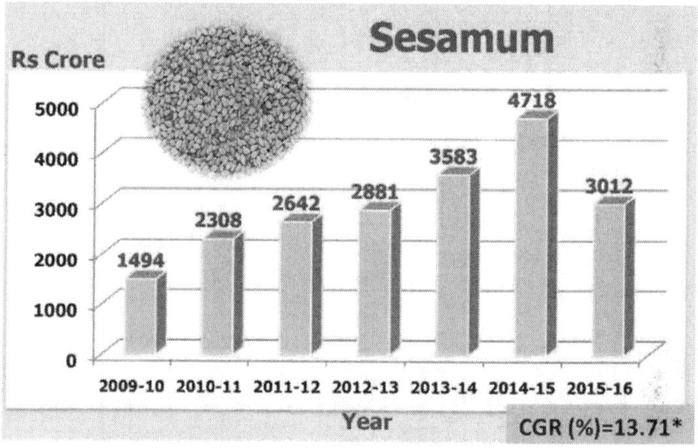

Fig. 5.9 Export of sesamum from India

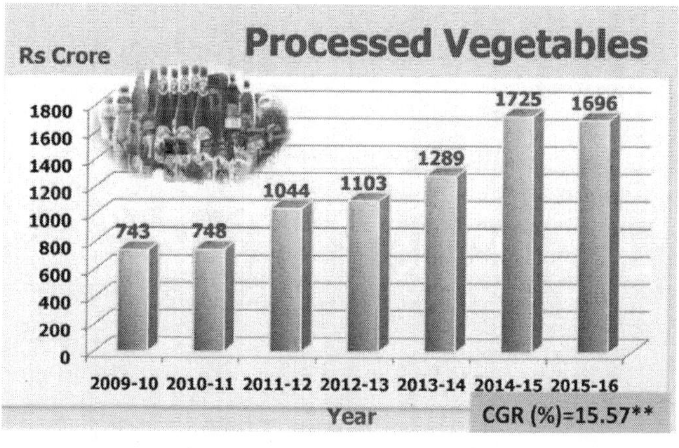

Fig. 5.10 Export of processed vegetables from India

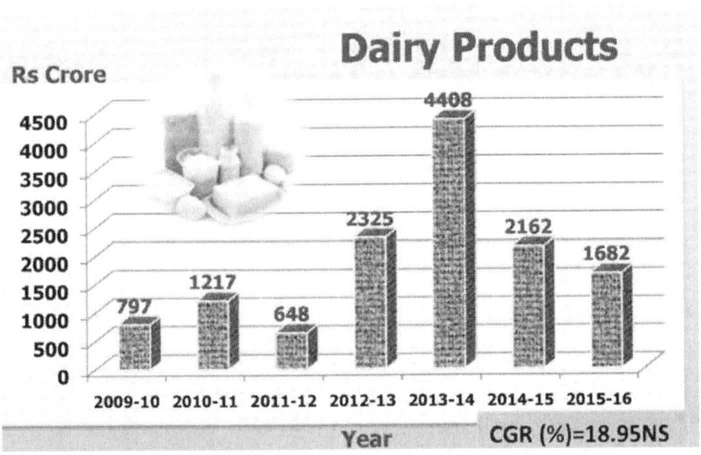

Fig. 5.11 Export of dairy products from India

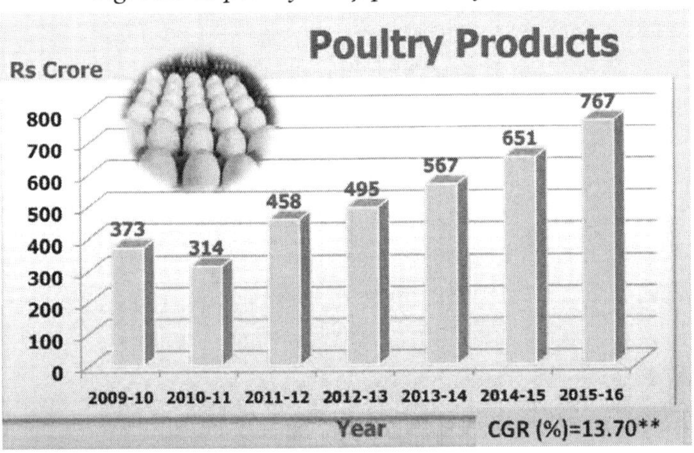

Fig. 5.12 Export of poultry products from India

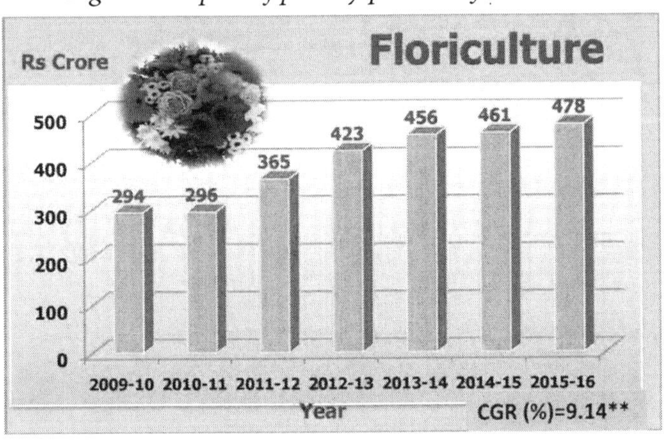

Fig. 5.13 Export of floriculture from India

6

EXPORT POTENTIAL OF FRUITS AND VEGETABLES

India's diverse climate ensures availability of all varieties of fresh fruits & vegetables. It ranks second in fruits and vegetables production in the world, after China. India produces nearly 15 and 11 per cent of the world's fruits and vegetables, respectively.

As per National Horticulture Database, India produced 86.60 million metric tonnes of fruits from 6.11 million hectares and 169.48 million metric tonnes of vegetables from 9.54 million hectares during 2014-15 (NHB, 2016). The state of Gujarat contributes 9.00 and 7.10 per cent in Indian fruits and vegetables production, respectively (NHB, 2015).

India is the largest producer of mango, banana, papaya and guava. Amongst vegetables, India is the largest producer of ginger and okra and ranks second in production of potato, onion, cauliflower, brinjal, cabbage, etc. The vast production base offers India tremendous opportunities for export.

Recently, the export of fruits and vegetables from India is slowly gaining pace. This has occurred due to concurrent developments in the areas of cold chain infrastructure, increase in productivity and quality through research, modern post-harvest technologies, favourable government policies and various initiatives taken by APEDA.

The fruits and vegetables are exported in fresh/chilled as well as in processed forms. Category-wise export of fruits and vegetables from India by quantity as well as by value were worked out. The results are presented in Table 6.1 and Fig 6.1 to 6.3.

The data presented in Table 6.1 shows that during the year 2015-16, India exported 3135 thousand tons of fruits and vegetables worth of Rs 14495 crore in various form viz., fresh fruits (504 thousand tons worth of Rs 3525 crore), processed fruits (450 thousand tons worth of Rs 3697 crore), fresh vegetables (1901 thousand tons worth of Rs 4867 crore), processed vegetables (269 thousand tons worth of Rs 1913 crore) and fruits/vegetables seeds (11 thousand tons worth of Rs 494 crore).

Looking to % share in total export of fruits and vegetables, fresh vegetable accounted for the highest share (33.58%) followed by processed fruits (25.50%), fresh fruits (24.32%) and processed vegetables (13.20%) whereas share of fruits/ vegetable seeds was very negligible (3.40%).

Table 6.1 Category-wise export of fruits and vegetables from India

Year	Fruits		Vegetables		Fruits/ Veg. Seeds	Total
	Fresh	Processed	Fresh	Processed		
Quantity (thousand tons)						
2006-07	353	330	1656	264	8	2612
2007-08	366	351	1361	253	10	2341
2008-09	470	368	2178	340	9	3365
2009-10	475	410	2088	299	9	3280
2010-11	418	370	1682	258	12	2739
2011-12	448	425	2044	323	15	3256
2012-13	498	417	2435	307	17	3674
2013-14	481	462	2436	275	18	3672
2014-15	427	471	2074	315	12	3299
2015-16	504	450	1901	269	11	3135
CGR%	2.79*	3.74**	3.28	0.57	5.90*	3.03
Value (Rs crore)						
2006-07	899	1168	1599	721	122	4508
2007-08	919	1223	1529	681	142	4494
2008-09	1163	1638	2514	993	120	6429
2009-10	1475	1685	3059	1043	145	7407
2010-11	1271	1816	2701	889	185	6861
2011-12	1799	2198	3033	1272	288	8591
2012-13	2504	2342	3483	1495	348	10171

Year	Fruits		Vegetables		Fruits/ Veg. Seeds	Total
	Fresh	Processed	Fresh	Processed		
2013-14	3298	3040	5463	1698	411	13909
2014-15	2771	3411	4703	2050	427	13362
2015-16	3525	3697	4867	1913	494	14495
CGR%	16.47**	13.22**	13.50**	12.50**	17.89**	14.04**
% to total (2015-16)	24.32	25.50	33.58	13.20	3.40	100.00

* *Significant at 5% level* ** *Significant at 1% level*

Further, it was observed that the export of fruits in both the categories viz. fresh and processed reported significant annual growth of 2.79 and 3.74%, respectively by quantity during the last decade.

Similarly, by value also they reported significant rise of 16.4 and 13.22% per annum, respectively. These trends show that there is a good scope for export of fresh as well as processed fruits from India.

It also indicates that processed fruits are moving at a faster rate as compare to fresh fruits. In case of vegetables, both the categories viz. fresh and processed did not show any significant growth by quantity.

But by value, both the categories reported significant growth of 13.50 and 14.50% per annum, respectively. This rise might be due to rise in prices of vegetables and also due to rise in value of US dollar during this period.

The export of fruit/vegetable seeds showed significant rise by quantity (5.90% per annum) as well as by value (17.89% per annum) during last decade.

Overall, export of fruits and vegetables by quantity did not show any significant growth but by value it reported a significant rise of 14.04% per annum.

The results indicate that export of fruits has peformed well but in case of vegetables, more efforts are needed in a more comprehensive manner along with appropreate government policy measures to improve our exports in future.

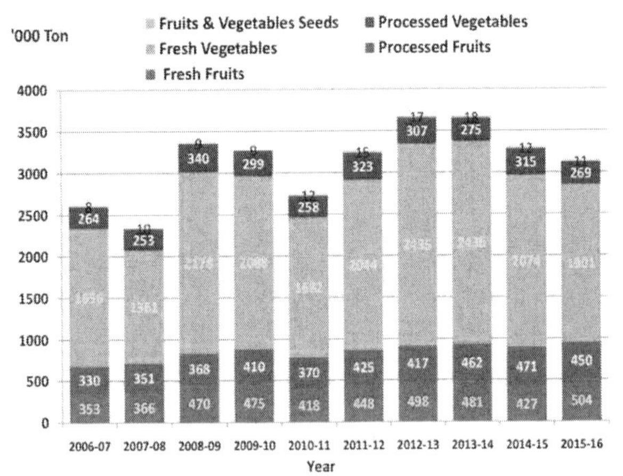

Fig. 6.1 Category-wise export of fruits and vegetables from India (by quantity)

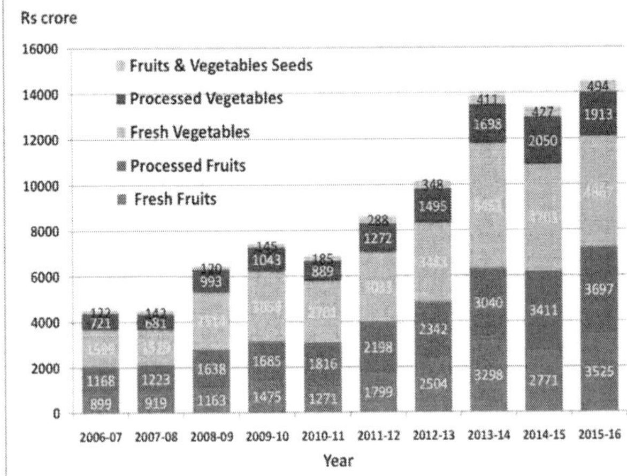

Fig. 6.2 Category-wise export of fruits and vegetables from India (by value)

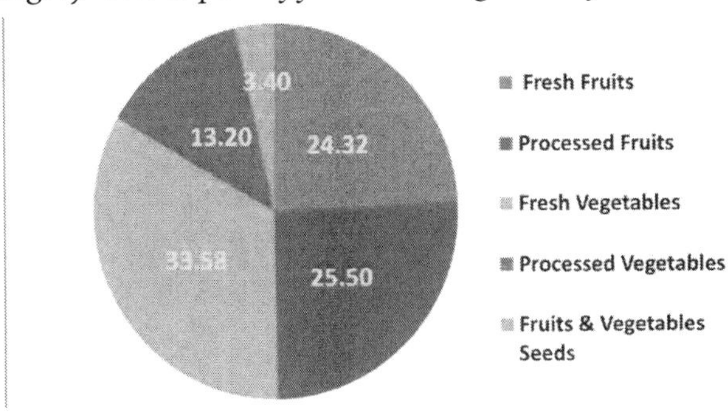

Fig. 6.3 Category-wise % share in total exports of fruits and vegetables from India (2015-16)

AGRICULTURE EXPORT POLICY

Department of Commerce has brought out a comprehensive "Agriculture Export Policy" aimed at doubling the agricultural exports and integrating Indian farmers and agricultural products with the global value chains. The Agriculture Export Policy has the following vision. "Harness export potential of Indian agriculture, through suitable policy instruments, to make India global power in agriculture and raise farmers' income."

Objectives of the Agriculture Export Policy are as under:

To diversify our export basket, destinations and boost high value and value added agricultural exports including focus on perishables.

To promote novel, indigenous, organic, ethnic, traditional and nontraditional Agri products exports.

To provide an institutional mechanism for pursuing market access, tackling barriers and deal with sanitary and phytosanitary issues.

To strive to double India's share in world agri exports by integrating with global value chain at the earliest.

Enable farmers to get benefit of export opportunities in overseas market.

Actions taken to implement AEP

The first ever Agriculture Export Policy (AEP) was introduced by the Government in December 2018.

As a part of the process of implementation of AEP, eighteen States viz. Maharashtra, U.P., Kerala, Nagaland, Tamil Nadu, Assam, Punjab, Karnataka, Gujarat, Rajasthan, Madhya Pradesh, Andhra Pradesh, Telangana, Manipur, Sikkim, Nagaland, Mizoram and Uttarakhand and the 2 UTs viz Ladakh and Andaman & Nicobar Islands have finalized the State specific Action Plan.

The State Level Monitoring committee (SLMC) has been formed in 25 States and 4 UTs. 28 States & 4 UTs have nominated respective Nodal agencies for implementation of AEP.

Cluster Development

As part of the Agriculture Export Policy, 46 unique product-district clusters have been identified for export promotion 29 Cluster Level Committees have been formed in cluster districts of different clusters.

Cluster activation for export: DoC intervened through APEDA for linking FPOs & exporters for activation of clusters. After the said linking, transportation/

logistics issues were solved and exports took place from land locked clusters. Following are some of the success stories:

Varanasi cluster (Fresh Vegetables): Till date, 48 MT of Fresh Vegetables (green chilly, long guard, green peas & cucumber), 10 MT of Mangoes (Banarasi, Langra, Ramkheda &Chausa) and 532 MT of Black rice have been exported from the cluster through FPOs.

Ananthpur Cluster (Banana): During the recent season (Jan-April 2021), 30,291 MT of banana have been dispatched from Ananthpur in Andhra Pradesh through 9 reefer rail movements and exported to Middle East.

Nagpur cluster (Orange): 115 MT of Nagpur Oranges and 45 MT of Ambiyabahar season Oranges have been exported to Middle East Countries by sea (for the first time) and supplied to top supermarkets i.e. Lulu Super mart, Safari Mall, Nesto, etc.

Lucknow cluster (Mango): 80.25 MT of mangoes (Dasheri, Langra, and Bombay green) have been exported from Lucknow cluster to Middle East countries. Theni cluster (Banana): During the last one year till date, 2400 MT of Cavendish and 1560 MT of G9 & Nendran banana have been exported from the cluster.

Pomegranate Cluster, Maharashtra – The export of Pomegranates from Solapur cluster district was 32,315 MT during the year 2020-21. Three FPOs formed by NABARD in the district are working for promoting exports.

Mango Cluster, Andhra Pradesh – In the current season, a consignment of Banganapalli (GI certified) & Survarnarekha mangoes (2.5 MTs) sourced from farmers in Krishna & Chittoor cluster districts was exported to South Korea on 06, May, 2021. Total 109 MT mangoes was exported to Middle East, EU, UK, and New Zealand from the cluster. During this mango season, a total quantity of 4000MT of mangoes was transported from Krishna cluster district, Andhra Pradesh to Delhi by rail.

Mango Cluster, Telangana – So far, more than 100 MT of fresh mangoes have been exported to EU, UK, Middle East.

Rose Onion cluster, Karnataka – Around 7168 MT of Rose onions have been exported from the Cluster to Malaysia, Singapore, Indonesia, Bangladesh and Sri Lanka from October 2020 to December 2020.

Banana Cluster, Gujarat – During April 2020 till date, 6198.26 MT of fresh banana has been exported to Middle East countries viz. Bahrain, Dubai, Georgia,

Iran, Oman, Saudi Arabia, Turkey, UAE, Iraq etc. from the cluster comprising Surat, Narmada & Bharuch.

Banana Cluster, Maharashtra – During 2020-21, 3278, 280 & 90 containers of Banana have been exported from Solapur, Jalgaon & Kolhapur respectively.

Onion Cluster, Maharashtra – During January to 15April 2021, 10,697 MT of fresh Onions has been exported from Nasik cluster to multiple destinations in SE Asia, Middle East, Bangladesh. Fourteen FPOs formed by NABARD in the district are working for promoting exports.

Grapes Cluster, Maharashtra – During 2020-21, so far, 6797 containers of 91,762 MT of fresh grapes have been exported to EU from Nasik cluster district. 1013 containers of 13,884 MT and one container of Raisin have been exported from the cluster district of Sangli to EU and other countries.

These clusters have been activated by utilizing the existing resources, with little or no additional investments. Exports from these clusters are taking place on regular basis.

Country specific agri export strategy reports: An interaction was carried out with 60 Indian Missions and stakeholders to identify the products, their potential and way forward for preparing the country-wise agri-export strategy. The information was compiled, a SWOT analysis was done and the missions were asked to vet the draft reports. The reports have been shared with various stakeholders.

Report on product-specific measures: A detailed analysis was carried out to address the requirements of SPS/TBT issues existing in trade, for boosting exports of India. The report titled "Tariff Disadvantages of India's Agricultural Exports" is based on the potential export products identified for boosting exports under Agriculture Export Policy. The report contains following information:-

Applicable Tariff rates of Indian agriculture export products into importing countries.

Tariff duty for India's Competitor into Importing Countries for Agriculture exports.

Tariff disadvantages in the potential market for India in agriculture exports.

Key actionable points

Virtual Buyer Seller Meets – A series of V-BSMs (Twenty four V-BSMs) have been organized with UAE, Kuwait, Indonesia, Switzerland, Belgium, Iran, Canada (Organic products), UAE & USA (GI products), Germany, South Africa, Australia,

Thailand, Oman, Bhutan, Azerbaijan, Qatar, Saudi Arabia, Nepal, Uzbekistan, Vietnam, Netherlands, Brunei and Cambodia (Animal products). E-catalogues were released while each BSM having details of participant exporters, importers, Trade associations.

Virtual Trade Fair (VTF) – APEDA took an initiative to develop its own Virtual Trade Fair (VTF) application, and same has been developed. The virtual platform will provide an opportunity for agri importers of several countries and our exporters to interact through participation. The first Virtual Trade Fair was organized for Cereals product sector during 10-12th March 2021. The next VTF for Fresh fruits and vegetables was held from 27-29th May2021.

Agri Cells at different Embassies of India-APEDA is consulting 13 Agri Cells in our Missions in different countries seeking inputs on real-time basis to further strengthen the existing Market Intelligence cell. The consolidated reports received from Agri Cells are being referred to while preparing the strategy pertaining to the specific countries.

Farmer Connect Portal: A Farmer Connect Portal has been set up on APEDA's website for providing a platform for FPOs/FPCs, cooperatives to interact with exporters. Around 2360 FPO/FPCs and 2324 exporters have been registered in the portal so far.

7

EXPORT OF FRUITS FROM INDIA

India's diverse climate ensures availability of all varieties of fresh fruits & vegetables. It ranks second in fruits and vegetables production in the world, after China. As per National Horticulture Database (Second Advance Estimates) published by National Horticulture Board, during 2019-20, India produced 99.07 million metric tonnes of fruits and 191.77 million metric tonnes of vegetables. The area under cultivation of fruits stood at 6.66 million hectares while vegetables were cultivated at 10.35 million hectares. According to FAO (2019), India is the largest producer of ginger and okra amongst vegetables and ranks second in production of potatoes, onions, cauliflowers, brinjal, Cabbages, etc. Amongst fruits, the country ranks first in production of Bananas (26.08%), Papayas (44.05%) and Mangoes (including mangosteens and guavas) (45.89%).

The vast production base offers India tremendous opportunities for export. During 2020-21, India exported fruits and vegetables worth Rs. 9,940.95 crores/ 1,342.14 USD Millions which comprised of fruits worth Rs. 4,971.22 crores/ 674.53 USD Millions and vegetables worth Rs. 4,969.73 crores/ 667.61 USD Millions.

Grapes, Pomegranates, Mangoes, Bananas, Oranges account for larger portion of fruits exported from the country while Onions, Mixed Vegetables, Potatoes, Tomatoes, and Green Chilly contribute largely to the vegetable export basket.

The major destinations for Indian fruits and vegetables are Bangladesh, UAE, Netherland, Nepal, Malaysia, UK, Sri Lanka, Oman and Qatar.

Though India's share in the global market is still nearly 1% only, there is increasing acceptance of horticulture produce from the country. This has occurred due to concurrent developments in the areas of state-of-the-art cold chain infrastructure and quality assurance measures. Apart from large investment pumped in by the private sector, public sector has also taken initiatives and with APEDA's assistance several Centers for Perishable Cargoes and integrated post harvest handling facilities have been set up in the country. Capacity building initiatives at the farmers, processors and exporters' levels has also contributed towards this effort.

AREA, PRODUCTION AND PRODUCTIVITY OF FRUITS

Fresh fruits export from India has witnessed a major growth. The major destinations of exports of fresh fruits include the Netherlands, Bangladesh, the UK, the UAE, Iran, Nepal, Saudi Arabia, Russia, Qatar, and Oman.

In April-January 2021-22, guava exports have risen to USD 2.09 million against USD 0.58 million recorded in April-January 2013-14.

Paneer (Indian cottage cheese) and Curd (yogurt) exports have increased as well, to USD 30 million in April–January 2021-22 from USD 10 million in April–January 2013-14.

Pomegranates, Grapes, Bananas, Mangoes, and Oranges are the most variety of fruits that are exported from the country. In vegetables, Mixed Vegetables, Onions, Tomatoes, Potatoes, and Green Chilly take the major share of exports from India.

The UAE, the United States, Bangladesh, Singapore, Bhutan, Malaysia, Saudi Arabia, Oman, Qatar, and Indonesia are all the major destinations of dairy export from India.

New restrictions, such as Russia's exclusion from the SWIFT system, are likely to have an impact on India's grape and mango exports. The restrictions will severely limit the ability of the Indian exporters to explore market opportunities in the United States. Currently, grapes are being exported in full swing from India and the sanctions on Russia will hamper this. The sanctions will not only affect the payments of the Indian exporters but will also hamper the movement of containers.

India is the second largest producer of fruits in the world after China and is known as fruit basket of world. India is endowed with a remarkably heterogeneous area characterized by a great diversity of agro climatic zones, allowing for production of a variety of horticultural crops such as fruits, vegetables, flowers, spices, plantation crops, root and tuber crops, and medicinal and aromatic crops.

The data in Table 7.1 shows, that area and production of horticultural crops is increased from 16.6 million ha and 145.8 million MT to 24.2 million ha and 277.4 million MT during the last 14 years. During the last two decades, the production became almost double in horticultural crops, which shows great potential in terms of income and foreign exchange earnings.

Table 7.1 Area, production and productivity of horticultural crops in India

Year	Area (Million ha)	Production (Million MT)	Productivity (MT/ha)
1991-92	12.8	96.6	7.5
2001-02	16.6	145.8	8.8
2002-03	16.3	144.4	8.9
2003-04	19.2	153.3	8.0
2004-05	21.1	170.8	8.1
2005-06	18.7	182.8	9.8
2006-07	19.4	191.8	9.9
2007-08	20.2	211.2	10.5
2008-09	20.7	214.7	10.4
2009-10	20.9	223.1	10.7
2010-11	21.8	240.5	11.0
2011-12	23.2	257.3	11.1
2012-13	23.7	268.8	11.3
2013-14	24.2	277.4	11.5

Source: Department of Agriculture, Cooperation and Farmers Welfare, GoI

The production share of fruits was 32.08 per cent among all horticultural crops, during the year 2013-14. The production of fruit crops became more than double in the last recent 14 years from 43 million MT to 88.98 million MT. During the year 2013-14, the area and production under fruit crops was 7.22 million ha and 88.98 million MT.

Table 7.2 Area, production and productivity of fruits in India

Year	Area ('000ha)	Production ('000 MT)	Productivity (MT/ ha)
1991-92	2874	28632	10.0
2001-02	4010	43001	10.7
2002-03	3788	45203	11.9
2003-04	4661	45942	9.9
2004-05	5049	50867	10.1
2005-06	5324	55356	10.4
2006-07	5554	59563	10.7
2007-08	5857	65587	11.2
2008-09	6101	68466	11.2
2009-10	6329	71516	11.3
2010-11	6383	74878	11.7
2011-12	6705	76424	11.4
2012-13	6982	81285	11.6
2013-14	7216	88977	12.3
% share to horticultural crops (2013-14)	29.82	32.08	

Source: Department of Agriculture, Cooperation and Farmers Welfare, GoI

The major fruits grown in India are Mango (35.41% of total area under fruits), Citrus (15.60%) and Banana (13.45%). Other grown fruits are Apple, Guava, Pomegranate, Grapes, Jackfruit, Pineapple, Papaya, Sapota, Aonla and others. Amongst fruits, the country ranks first in production of Banana (33.74%) followed by Mango (21.39%) and citrus (13.46%).

Table 7.3 Area and production of fruits in India (2014-15)

Fruits	Area			Production		
	'000 ha	% share	Rank	'000 MT	%	Rank
Mango	2163	35.41	1	18527	21.39	2
Citrus	953	15.60	2	11655	13.46	3
Banana	822	13.45	3	29221	33.74	1
Apple	319	5.22	4	2134	2.46	7
Guava	246	4.03	5	3994	4.61	5
Pomegranate	181	2.96	6	1789	2.07	10
Grapes	123	2.01	7	2823	3.26	6
Jackfruit	118	1.94	8	2088	2.41	8
Pineapple	116	1.90	9	1984	2.29	9
Papaya	115	1.88	10	4913	5.67	4
Sapota	106	1.74	11	1339	1.55	11
Aonla/Gooseberry	95	1.56	12	1173	1.35	12
Others	751	12.29	-	4962	5.73	-
Total Fruits	6110	100.00		86602	100.00	

Source: NHB

Table 7.4 Major fruits producing states in India (2013-14)

States	Area ('000 ha)	Production ('000 MT)
Maharashtra	1565.0	13457.9
Andhra Pradesh	640.1	10510.6
Gujarat	370.8	8002.0
Tamil Nadu	328.6	7369.9
Uttar Pradesh	379.0	6887.4
Karnataka	396.0	6652.4
Madhya Pradesh	203.8	5696.0
Telangana	364.5	4441.0
Bihar	302.1	4013.6
West Bengal	223.5	2909.7
Others	2442.6	19036.5

Source: NHB

The major producing states in the country are Maharashtra, Andhra Pradesh, Gujarat, Tamil Nadu, Uttar Pradesh, Karnataka, Madhya Pradesh, Telangana, Bihar, West Bengal and others.

Maharashtra ranks first in terms of area under fruits in India followed by Andhra Pradesh, Karnataka, Uttar Pradesh and Gujarat. Gujarat State ranks 5th in terms of area (5% share to India's area) and 3rd in terms of production (9% share to India's production), during the year 2013-14.

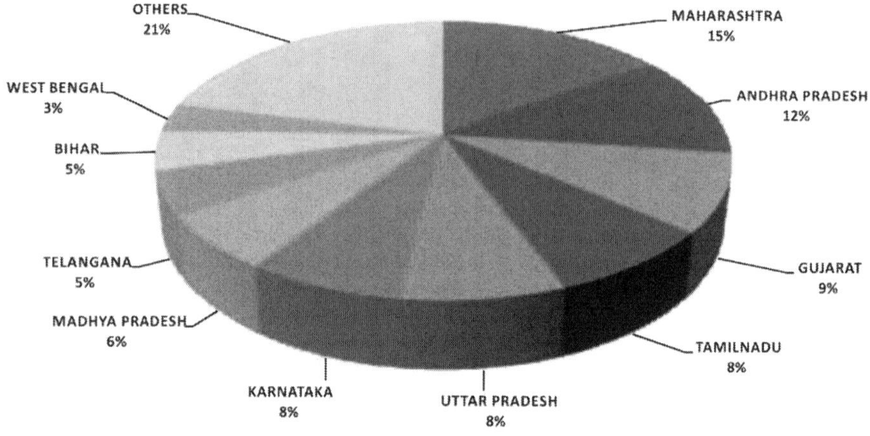

Fig. 7.1 Leading fruits producing states in India (2013-14)

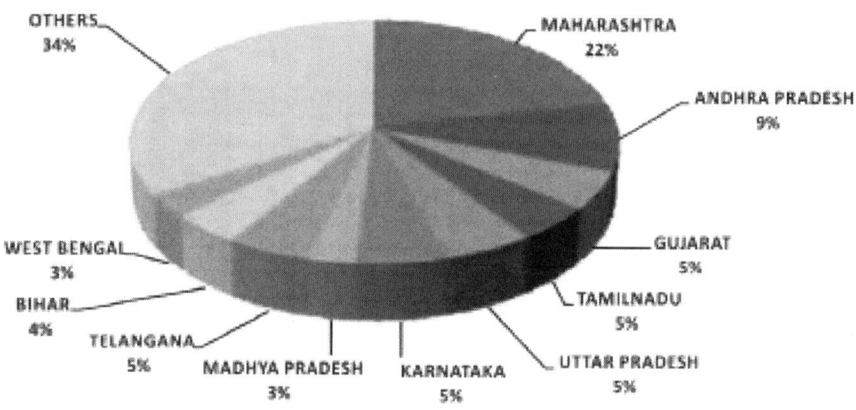

Fig. 7.2 Leading fruits growing states in India (2013-14)

During the year 2014-15, area and production in Gujarat state was 0.39 million ha and 8.33 MT respectively. Amongst all fruits, Mango ranks first in area with 38.20% share and second in production with 14.65% share. Banana is the major producing fruit crop of Gujarat state.

Table 7.5 Area and production of fruits in Gujarat (2014-15)

Sr. No.	Fruits	Area (ha)	% to Total	Production (MT)	% to Total
1	Mango	150048	38.20	1219710	14.65
2	Banana	67016	17.06	4324357	51.92
3	Citrus	41742	10.63	462416	5.55
4	Chiku	29421	7.49	321318	3.86
5	Coconut(1000 nuts)*	22451	-	211700	-
6	Papaya	19128	4.87	1170638	14.06
7	Date Palm	17882	4.55	164282	1.97
8	Pomegranate	14768	3.76	171659	2.06
9	Ber	11880	3.02	120916	1.45
10	Guava	11144	2.84	146058	1.75
11	Aonla	9669	2.46	95624	1.15
12	Cashew Nut	8422	2.14	27688	0.33
13	Custard Apple	5336	1.36	55041	0.66
14	Other	6390	1.63	48597	0.58
	Total	392846	100.00	8328302	100.00

* not included in total Source : Agriculture and Cooperation Department, GoG

Product-wise export of major fruits

India is also a major exporter of fruits to the world. The vast production base offers India tremendous opportunities for export. During 2015-16, India exported fruits and vegetables worth Rs 8,391.41 crores which comprised of fruits worth Rs 3,524.50 crores. Mangoe, walnut, grape, banana and pomegranates accounts for larger portion of fruits exported from the country. Fruit have a great demand in the international market. Therefore, they are one of the potential earners of foreign exchange.

The total fruits (fresh + processed) export in terms of value was 7221 crores during the year 2015-16. The product-wise export of major fruits during the year 2015-16 is presented in Table 7.6 and Fig 7.3. It shows that mango ranked first with Rs 1961.93 crore accounting for 27.17% in the total export of fruits from India followed by grapes (21.35%). These two fruits i.e. mango and grape together contributes about half of the fruits export. Whereas pomegranate and banana contributes 5.76 and 4.93% respectively.

Table 7.6 Product-wise export of major fruits (2015-16)

Rank	Fruit Crop	Product	Value in Rs Crore		% share
1	Mango	Fresh/Dried	317.10	1961.93	27.17
		Pulp	796.17		
		Squash	61.02		
		Jams/Jellies/ Marmalades	742.82		
		Sliced Dried	44.82		
2	Grape	Fresh	1337.78	1541.95	21.35
		Dried (Raisins)	204.17		
3	Pomegranate	Fresh	-	416.00	5.76
4	Banana	Fresh	-	356.13	4.93
5	Walnut	Fresh	-	117.92	1.63
6	Citrus	Lemons & Limes Fresh/Dried	74.07	86.81	1.20
		Other Citrus Fruit Fresh/Dried	12.74		
7	Melon	Water Melons, Fresh	54.46	80.55	1.12
		Other Melons Fresh	26.09		
8	Papaya	Fresh/Dried	-	56.78	0.79
9	Guava	Guavas Fresh/ Dried	7.12	33.32	0.46
		Jams/Jellies/ Marmalades	26.20		
10	Sapota	Fresh	-	9.67	0.13
	Other	-	-	2559.94	35.45
	Total (Fresh + Processed)			7221.00	100.00

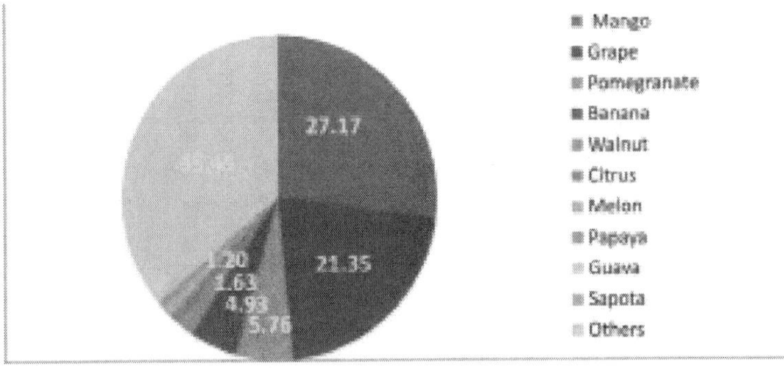

Fig 7.3 Commodity-wise share (%) in total export of fruits (2015-16)

Year	Guava		Grapes	
	Quantity	Value	Quantity	Value
2016-17	1,408.16	0.91	1,98,471.30	267.04
2017-18	1,229.77	0.86	2,05,039.41	303.71
2018-19	978.69	0.67	2,53,619.00	335.11
2019-20	1,697.14	0.73	1,96,376.51	303.46
2020-21	2,886.37	1.27	2,47,187.05	314.11

India is popularly known as 'the fruits basket of the world'. In India, wehave been blessed with favorable climatic and soil conditions which help the growth of not only fruits in India, but also vegetables, grains, pulses, and such other crops. Most of the fruits are grown in India including mangoes, bananas, papayas, oranges, apricots, grapes, strawberries, apples, guavas, litchis, so on, and so forth. Therefore, fruits export from India has been on the rise for a few decades. The export of fruits from India was around 372213.73 Metric tonnes in the year 2018-2019.There are numerous fruits importer from India including countries like Bangladesh, Nepal, Oman, Saudi Arabia, UAE, and many more. Pisum Foods us one of the leading fruits exporter in India. We conduct export of Indian fruits to various parts of the world. Being one of the most experienced fruits exporters, our motto is to spread the Indian food products and the Indian culture to every corner of the world. We at Pisum Foods make sure that your food products reach the buyer well packaged and on the promised time. If you are looking for fruits import from India, Pisum can help you at every step on the way!

The export of fresh fruits from India has increased from USD 516.26 million in 2014-15 to USD 768.54 million in 2020-21.

Details of India's exports of guava and grapes during the last five years are as under:

Quantity in MT; Value in USD million

Source: DGCI&S

Promotion of exports of agriculture products, including dairy products like curd and cheese, is a continuous process. To promote the export of dairy products, an Export Promotion Forum (EPF) has been constituted under the aegis of Agricultural & Processed Food Products Export Development Authority (APEDA). The EPF has representation from trade/industry, line Ministries/ Departments, regulatory agencies, research institutes, state governments etc. Meetings of the EPF are held regularly to discuss the various issues affecting exports, such as SPS/ TBT issues, market access issues, plans for export promotion and capacity building programmes etc. Recommendations made by the EPFs are passed on to the relevant authorities for appropriate action.

Mathura and Banaskantha districts in Uttar Pradesh and Gujarat respectively have been identified as clusters for export of dairy products. APEDA has organized awareness cum capacity-building programmes for farmers/Farmer-Producer Organisations (FPOs) in these clusters. APEDA also provides assistance to exporters of dairy products under various components of its export promotion scheme.

Mango

Mango, the word itself keeps us tempting. With the onset of the summer season itself, kids start demanding for ripe and juicy mangoes. Not just kids but everyone loves to relish mangoes. It is the "King of Fruits". Also referred to as the "peach of the tropics," this fruit is one of the main types of fruit considered exotic.

Because it is sweet, refreshing and aromatic, it is commercialized throughout the world. Amongst all fruits, Mango ranks first in area with 35.41% share and second in production with 21.39% share in India.

Mango is commercially grown in more than 80 countries. India is the major Mango growing country, contributing nearly 46.74 per cent of world's area and 40.48 per cent of world's production. In India, during the year 2014-15, area under mango was 2.16 million ha and the production was 18.53 million MT (Table 7.7).

Table 7.7 Area, production and productivity of mango in India

Sr. No	Year	Area ('000 ha)	Production ('000 MT)	Productivity (MT/ ha)
1	2005-06	2081	12663	6.1
2	2006-07	2154	13734	6.4
3	2007-08	2201	13997	6.4
4	2008-09	2309	12750	5.5
5	2009-10	2312	15027	6.5
6	2010-11	2297	15188	6.6
7	2011-12	2378	16196	6.8
8	2012-13	2500	18002	7.2
9	2013-14	2516	18431	7.3
10	2014-15	2163	18527	8.6

Source: NHB

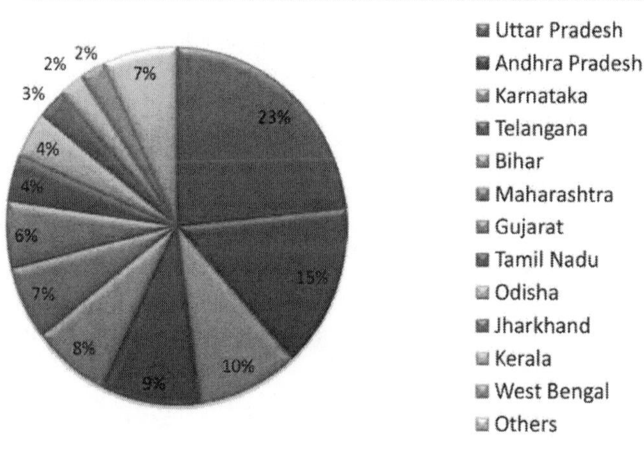

Fig 7.4 Leading mango producing states in India (2013-14)

The major mango growing states are Andhra Pradesh, Uttar Pradesh, Karnataka, Bihar, Gujarat and Tamil Nadu. Uttar Pradesh is the leading producing state of mango, contributes 23% of total production in India, followed by Andhra Pradesh (15%) and Karnataka (10%) (Fig 7.4).

Mangoes are exported in various forms viz., fresh/dried, pulp squash, jams/ jellies/ marmalades and slice. The growth in exports of mango during study period was worked out and presented in Table 7.8.

The export of fresh/dried mango, in term of quantity, was significantly declined by 7.86% annually during the last decade whereas the export in form of Jams/Jellies/Marmalades has significantly increased by 15.83% annually during the same period. It indicates that the demand for mango shifted from fresh mango to processed products in export market. Other forms of mango exports viz., pulp, squash and sliced did not show any significant change.

In value term, export of all forms of mango except squash significantly increased during the last decade with an overall growth of 9.90% per annum. Further, it was observed that pulp and Jams/ Jellies/ Marmalades products were the major contributors in mango export, with a share of 40.58 and 37.86%, respectively during the year 2015-16, whereas contribution of squash and slice in total mango export was very limited.

Table 7.8 Export of mango from India

Year	Fresh/ Dried	Pulp	Squash	Jams/Jellies/ Marmalades	Sliced Dried	Total
Quantity (in tons)						
2006-07	79061	156836	1778	21086	2991	261752
2007-08	54351	166752	13022	31076	5045	270246
2008-09	83703	173014	5584	32906	1499	296705
2009-10	74461	186198	653	44902	2101	308314
2010-11	58863	170220	1439	44140	1210	275873
2011-12	63441	150499	5205	58090	5481	282717
2012-13	55585	147816	7709	62304	2265	275678
2013-14	41280	174860	8887	73868	2758	301653
2014-15	42998	154821	10304	88333	4193	300649
2015-16	36329	128866	9457	94511	3644	272807
CGR%	-7.86**	-1.85	14.80	15.83**	3.19	0.54

Year	Fresh/ Dried	Pulp	Squash	Jams/Jellies/ Marmalades	Sliced Dried	Total
Value (in Rs Crore)						
2006-07	141.94	505.83	6.87	93.6	10.95	759.19
2007-08	127.42	509.69	49.09	138.11	7.58	831.89
2008-09	170.71	752.99	31.83	207.22	4.83	1167.58
2009-10	200.54	744.61	2.92	245.89	7.81	1201.77
2010-11	164.84	818.93	5.91	236.27	8.4	1234.35
2011-12	209.74	620.83	25.37	339.56	19.01	1214.51
2012-13	264.72	608.56	41.59	388.46	17.73	1321.06
2013-14	285.43	772.95	54.89	555.28	26.86	1695.41
2014-15	302.54	841.39	63.21	622.16	38.81	1868.11
2015-16	317.1	796.17	61.02	742.82	44.82	1961.93
CGR%	10.26**	4.15*	20.35	21.72**	21.80**	9.90**
% to total (2015-16)	16.16	40.58	3.11	37.86	2.28	100.00

* Significant at 5% level ** Significant at 1% level

Top 10 Export Destinations for various forms of Mango during 2015-16 are reported in Table 7.9. It shows that Saudi Arab was the largest export market for pulp contributing about 27.88% of total pulp export of our country followed by Netherland (10.93%) and Yemen (6.88%). In case of Jams/Jellies/Marmalades, the top most export destination was Netherland (26.84%) followed by Iran (10.72%) and USA (8.23%).

Table 7.9 Top 10 export destinations for mango (2015-16)

Rank	Country	Value Rs Crore	% to Total	Rank	Country	Value Rs Crore	% to Total
	Pulp				Jams/Jellies/Marmalades		
1	Saudi Arab	221.97	27.88	1	Netherland	199.36	26.84
2	Netherland	87.01	10.93	2	Iran	79.65	10.72
3	Yemen	54.78	6.88	3	U S A	61.15	8.23
4	Kuwait	51.46	6.46	4	Saudi Arab	54.25	7.30

Rank	Country	Value Rs Crore	% to Total	Rank	Country	Value Rs Crore	% to Total
5	U A E	46.92	5.89	5	U K	52.76	7.10
6	U S A	40.06	5.03	6	U A E	38.74	5.22
7	U K	37.44	4.70	7	France	31.51	4.24
8	Sudan	34.65	4.35	8	Japan	31.07	4.18
9	Germany	22.89	2.88	9	China	21.50	2.89
10	China	20.52	2.58	10	Yemen	18.91	2.55
Total (Incl. others)		796.17	100.00	Total (Incl. others)		742.82	100.00
	Fresh/Dried				Squash		
1	U A E	191.99	60.55	1	Netherland	12.80	20.98
2	U K	32.06	10.11	2	Saudi Arab	9.90	16.22
3	Saudi Arab	16.75	5.28	3	Iran	9.58	15.71
4	Nepal	13.94	4.39	4	Nepal	5.69	9.32
5	Kuwait	12.98	4.09	5	U A E	3.72	6.10
6	Qatar	10.23	3.23	6	France	3.01	4.94
7	U S A	7.34	2.32	7	U K	2.20	3.61
8	Baharain	6.33	2.00	8	Sri Lanka	1.99	3.26
9	Singapore	6.26	1.97	9	U S A	1.63	2.67
10	Oman	4.13	1.30	10	Tunisia	1.41	2.30
Total (Incl. others)		317.10	100.00	Total (Incl. others)		61.02	100.00

Quality is a very important aspect in export of fruits and vegetables. Export specifications for fruits and vegetables vary from country to country. Such quality parameters required by major importing countries in Mango crop are presented as under table 7.10.

Table 7.10 Export specifications for mango

Particulars	Variety	Middle East	Netherlands /Germany	UK	Japan	USA
Weight	Alphanso	200-250 g	250-300 g	250-300 g	250-300 g	250-300 g
	Kesar	200-250 g	225-250 g	225-250 g	250-300 g	250-300 g

Particulars	Variety	Middle East	Netherlands /Germany	UK	Japan	USA
Packing		1 Dozen/ 2.5 kg	1 Dozen/ 2.5 kg	1Dozen/ 2.5 kg	1 Dozen / 3.5 kg	1 Dozen / 3.5 kg
Storage Temp.		13°c	13°c	13°c	13°c	13°c
Transportation		By Sea	By Air	By Air	By Air	By Air

Grape

Grape is one of the important fruits of India. In India, during the year 2013-14, area and production was 118.7 thousand ha and 2585.3 thousand MT, respectively. Area and production became almost double during the last decade in the country (Table 7.11). Major grape growing states are Maharashtra, Karnataka, Andhra Pradesh, Tamil Nadu, and the North-Western region covering Punjab, Haryana, Western Uttar Pradesh, Rajasthan and Madhya Pradesh. Maharashtra ranks first in terms of production accounting for more than 82.56% of total production and highest productivity in the country, followed by Karnataka (11.70%).

India is one of the major exporters of fresh Grape to the world. The details of export of grape from India are given in Table 7.12. It shows that the country has exported 155246 tons of grape worth of Rs 1541.95 crore to the world during the year 2015-16. Looking to the growth, fresh grape did not show any significant change in export by quantity but raisins (dried grape) reported 53.42% annual growth during last decade. By export value, both fresh as well as raisins reported 18.70 and 61.82% growth, respectively with an overall growth of 20.07% per annum. The fresh grapes, alone occupied 86.76% share in total grape export from India during the year 2015-16.

Table 7.11 Area, production and productivity of grape in India

Year	Area ('000 ha)	Production ('000 MT)	Productivity (MT/ha)
2004-05	60.5	1564.7	25.9
2005-06	66	1649.6	25
2006-07	65	1685	25.9
2007-08	68	1735	25.5
2008-09	80	1878	23.5
2009-10	106.4	880.7	8.3
2010-11	111	1235	11.1
2011-12	116	2220.9	19.1
2012-13	117.6	2483.1	21.1
2013-14	118.7	2585.3	21.8

Source: NHB

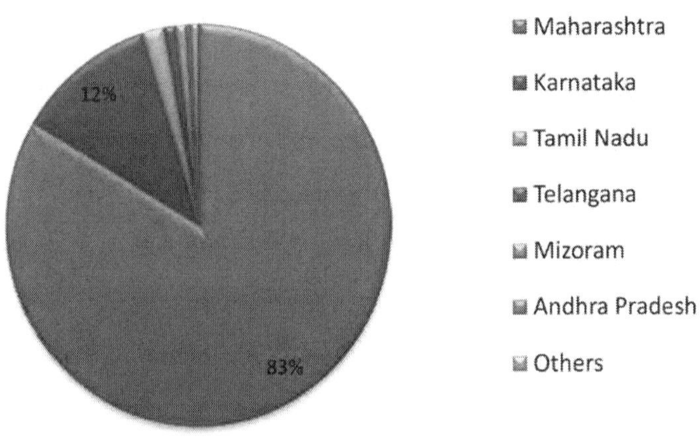

Fig 7.5 Leading grape producing states in India (2013-14)

Table 7.12 Export of grape from India

Year	Qty in Tons			Value in Rs Crore		
	Fresh	Raisins (Dried)	Total	Fresh	Raisins (Dried)	Total
2006-07	85563	218	85781	300.58	0.64	301.22
2007-08	96723	101	96824	317.07	0.46	317.53

Year	Qty in Tons			Value in Rs Crore		
	Fresh	Raisins (Dried)	Total	Fresh	Raisins (Dried)	Total
2008-09	118133	6143	124276	367.06	38.96	406.02
2009-10	117338	13476	130814	431.07	111.69	542.76
2010-11	92437	5233	97670	391.01	34.67	425.68
2011-12	94860	9542	104402	516.76	60.93	577.69
2012-13	140967	30042	171008	982.04	268.84	1250.88
2013-14	160256	31602	191859	1437.07	223.12	1660.19
2014-15	94377	12326	106703	972.77	107.64	1080.41
2015-16	128422	26825	155246	1337.78	204.17	1541.95
CGR%	3.38	53.42**	5.49	18.70**	61.82**	20.07**
% to total (2015-16)	-	-	-	86.76	13.24	100.00

** Significant at 1% level

The major export destinations for grape are given in Table 7.13. It shows that in case of fresh grape, Netherland ranked first with 43.80% share followed by UK (15.44%) and Russia (9.50%). While in case of raisins, Saudi Arab ranked first with 11.06% share followed by Ukrain (10.79%) and UAE (9.14%).

Table 7.13 Export specifications for grape

Particulars	Variety	Middle East	Holland/ Germany	UK
Berry Size/ colour	Thompson Seedless	15 mm Amber	16 mm White/ Amber	18 mm White
	Sharad Seedless	15 mm, Black	16 mm, Black	18 mm, Black
	Flame Seedless	-----	16 mm, Pink	18 mm, Pink
Packing		1 kg	4.5 kg /9 kg	4.5 kg /9 kg
Storage Temp.		0-1 °C	0-1 °C	0-1 °C

Table 7.14 Top 10 Export destinations for grape (2015-16)

Rank	Country	Value Rs Crore	% to Total	Rank	Country	Value Rs Crore	% to Total
Fresh				Raisins (Dried)			
1	Netherland	585.89	43.80	1	Saudi Arab	22.59	11.06
2	U K	206.54	15.44	2	Ukraine	22.04	10.79
3	Russia	127.05	9.50	3	U A E	18.66	9.14
4	U A E	97.94	7.32	4	Mexico	13.38	6.55
5	Germany	49.94	3.73	5	Poland	10.91	5.34
6	Saudi Arab	46.81	3.50	6	Malaysia	10.46	5.12
7	Thailand	40.12	3.00	7	Russia	9.57	4.69
8	Sri Lanka	22.85	1.71	8	Sri Lanka	9.23	4.52
9	Finland	20.13	1.50	9	Lithuania	8.63	4.23
10	Bangladesh	16.31	1.22	10	Romania	8.00	3.92
Total (Incl. others)		1337.78	100.00	Total (Incl. others)		204.17	100.00

Export quality specifications for Grape are presented in table 7.14. Quality parameters vary from country to country.

Pomegranate

Pomegranate is grown in tropical and subtropical regions of the world. It has been recognised internationally as super food considering its nutritional value.

During the year 2013-14, the area and production in the country was 130.77 thousand ha and 1345.72 thousand MT, respectivly. In India, Maharashtra is the leading producer of pomegranate, contributing 70.22 per cent to the total India's production followed by Karnataka (9.97%), Gujarat (7.37%) and Andhra Pradesh (6.68%), Gujarat and Tamil Nadu.

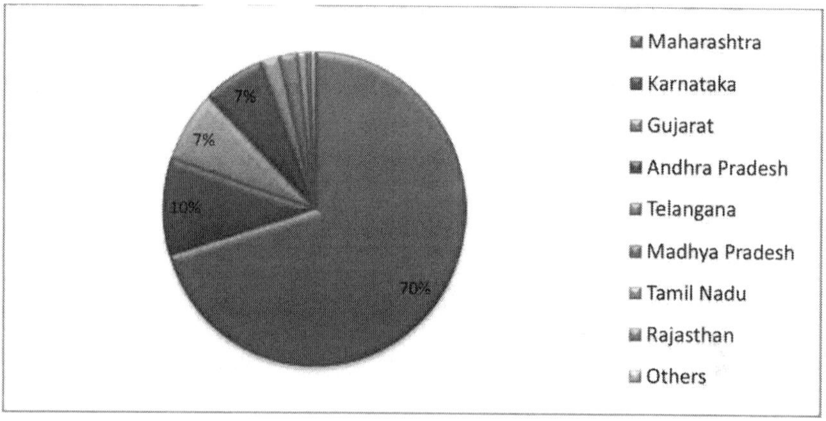

Fig. 7.6 Leading pomegranate producing states in India (2013-14)

Data on export of pomegranate (fresh) from India are presented in Table 7.15 which indicate that the export by quantity did not show any change but by value it reported significant growth of 18.97% per annum during last decade.

Table 7.15 Export of pomegranate (fresh) from India

Year	Qty in Tons	Value in Rs Crore
2006-07	21670	79.57
2007-08	35175	91.19
2008-09	34811	114.62
2009-10	33415	119.43
2010-11	18212	70.95
2011-12	30162	147.28
2012-13	36027	234.5
2013-14	31328	298.52
2014-15	20997	323.61
2015-16	31072	416.00
CGR%	-0.1	18.97**

** Significant at 1% level

The major export destinations for Indian pomegranate are presented. It was noticed that UAE alone accounted for 61.01% in the total export of pomegranate from India in the year 2015-16 followed by Saudi Arab (7.63%) and Netherland (5.12%).

Table 7.16 Top 10 export destinations for pomegranate (fresh) (2015-16)

Rank	Country	Value Rs Crore	% to Total
1	U A E	253.80	61.01
2	Saudi Arab	31.76	7.63
3	Netherland	21.30	5.12
4	U K	14.85	3.57
5	Thailand	11.09	2.67
6	Kuwait	10.74	2.58
7	U S A	8.72	2.10
8	Malaysia	8.71	2.09
9	Sri Lanka	6.05	1.45
10	Baharain	5.96	1.43
Total (Incl. others)		416.00	100.00

Export quality specifications for Pomegranate are presented in Quality parameters vary from country to country.

Table 7.17 Export specifications for pomegranate

Particulars	Variety	Middle East	Netherlands/ Germany	UK
Size/Weight/ Colour	Ganesh, Bhagwa	300-450 g Red	250-300 g Red	250-300 g Red
	Arakata, Mrudula	200-250 g Deep red	200-250 g Deep red	200-250 g Deep red
Packing		5 kg	3 kg	3 kg
Storage		5 °C	5 °C	5 °C
Transportation		By Sea	By Sea	By Sea

Banana

Banana is the world's favourite fruit and the fourth most important food crop after rice, wheat and maize. India is the world's largest producer of banana, with 28% share. In India, during the year 2013-14, area, production and yield was 802.6 thousand ha, 29.724 million MT and 37 MT/ha respectively. The major banana growing states are Tamil Nadu, Gujarat, Maharashtra, Andhra Pradesh and Karnataka. The major producing states in India are Tamil Nadu (19%), Maharashtra (16.25%) and Gujarat (14.22%).

Table 7.18 Area, production and productivity of banana in India

Year	Area ('000 ha)	Production ('000 MT)	Productivity (MT/ha)
2004-05	589.6	16744.5	28.4
2005-06	569.5	18887.8	33.2
2006-07	604	20998	34.8
2007-08	658	23823	36.2
2008-09	709	26217	37
2009-10	770.3	26469.5	34.4
2010-11	830	29780	35.9
2011-12	796.5	28455.1	35.7
2012-13	776	26509.1	34.2
2013-14	802.6	29724.6	37

Source: NHB

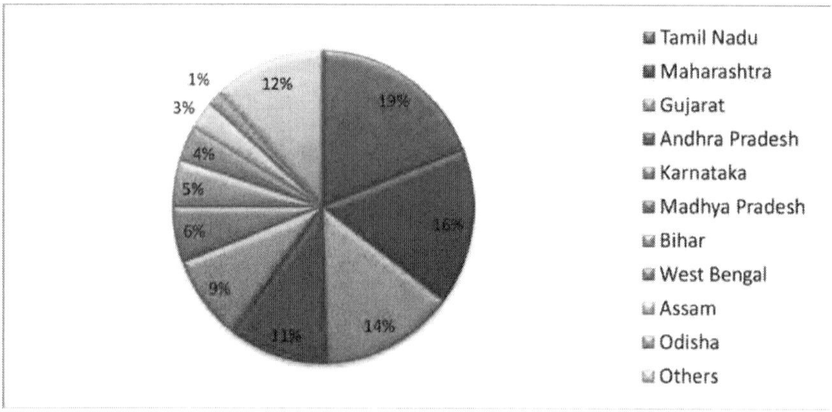

Fig 7.7 Leading banana producing states in India (2013-14)

Banana is the fourth largest contributor in total export of fruits from India. The details of banana export are presented in Table 4.19. It shows that export of banana was increased from 34831 tons in 2013-14 to 93673 tons in 2015-16. In value term also it increased from Rs 153.85 crore to Rs 356.13 crore during the same period. Due to insufficient time series data, growth rates could not be worked out.

Table 7.19 Export of banana (fresh) from India

Year	Qty in Tons	Value in Rs Crore
2013-14	34831	153.85
2014-15	63274	241.95
2015-16	93673	356.13

*Note: The export data was available for the year 2013-14 to 2015-16.

Major export destinations for Indian banana during 2015-16 are given in Table 7.20. UAE, Iran and Saudi Arab are leading the list contributing 32.93, 18.99 and 17.74% in total banana export from India, respectively.

Table 7.20 Top 10 export destinations for Indian banana (fresh) (2015-16)

Rank	Country	Value Rs Crore	% to Total
1	U A E	117.29	32.93
2	Iran	67.63	18.99
3	Saudi Arab	63.18	17.74
4	Oman	32.00	8.99

Rank	Country	Value Rs Crore	% to Total
5	Kuwait	24.70	6.94
6	Qatar	17.81	5.00
7	Nepal	16.37	4.60
8	Baharain	10.54	2.96
9	Maldives	4.51	1.27
10	Malaysia	0.37	0.10
Total (Incl. others)		356.13	100.00

Export quality specifications for banana are presented in Table 7.21. Quality parameters vary from country to country.

Table 7.21 Export specifications for banana

Particulars	Variety	Middle East
Size/Weight/ Colour	Grand Naine Cavendish	Colour : Green, Weight of Bunch : 2.5 kg Fruits preferably straight
Packing		13 kg
Storage		13-14 °C
Transport		By Sea

Walnut

Walnut is the most important temperate nut fruit of the country. During the year 2013-14, in India area and production under walnut was 121.87 thousand ha and 240.63 thousand MT. The major growing area is in Jammu and Kashmir, Uttaranchal, Himachal Pradesh and Arunachal Pradesh. Out of these states, Jammu & Kashmir occupying the largest share in total area and production.

The growth in export of walnut from India during last decade is given in Table 7.22. The country has exported 3292 tons of walnuts worth of Rs 117.92 crore to the world during the year 2015-16. But it did not show any significant growth during last decade.

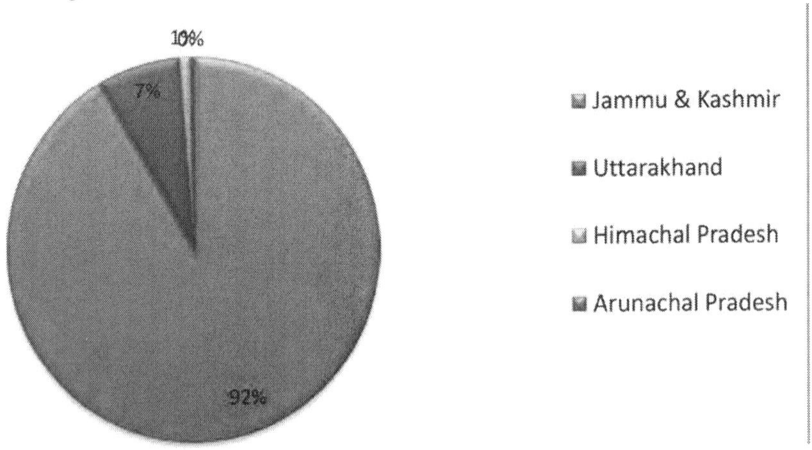

Fig 7.8 Leading walnut producing states in India (2013-14)

Table 7.22 Export of walnut from India

Year	Qty in Tons	Value in Rs Crore
2006-07	5063	118.04
2007-08	6716	162.08
2008-09	5696	141.24
2009-10	9073	197.90
2010-11	5762	166.29
2011-12	5842	231.08
2012-13	5295	199.84
2013-14	6726	324.53
2014-15	2666	136.45
2015-16	3292	117.92
CGR%	-6.73	2.00

Top 10 export destinations for walnut during 2015-16 are presented in Table 7.23. Amongst major export destinations, Germany stood first with 17.30% share followed by UK (14.36%), Netherlands (11.11%), Egypt (10.75%) and USA (9.66%).

Table 7.23 Top 10 export destinations for walnut (2015-16)

Rank	Country	Value Rs Crore	% to Total
1	Germany	20.40	17.30
2	UK	16.94	14.36
3	Netherland	13.11	11.11
4	Egypt	12.68	10.75
5	USA	11.39	9.66
6	France	9.09	7.71
7	Spain	7.81	6.63
8	Nepal	4.21	3.57
9	Belgium	3.98	3.38
10	Sweden	3.12	2.65
Total (Incl. others)		117.92	100.00

Citrus

Citrus comprises many species of economic importance. Citrus mainly includes Lime/lemon, mandarin and sweet orange (mosambi). India holds third position in citrus fruits production. Orange accounts for 50% of all citrus fruits production. In India, during the year 2013-14, area and production under citrus crop was 1077.7 thousand ha and 11147 thousand MT (Table 7.24). The production became almost double in last decade.

Table 7.24 Area, production and productivity of citrus in India

Year	Area ('000 ha)	Production ('000 MT)	Productivity (MT/ ha)
2004-05	708.2	5932.9	8.4
2005-06	742.3	6139	8.3
2006-07	798	7145	9
2007-08	867	8015	9.2
2008-09	923	8608	9.3
2009-10	987.3	9638	9.8
2010-11	846	7464	8.8
2011-12	915.1	7922.1	8.7
2012-13	1042.5	10089.7	9.7
2013-14	1077.7	11147.1	10.3

Source: NHB

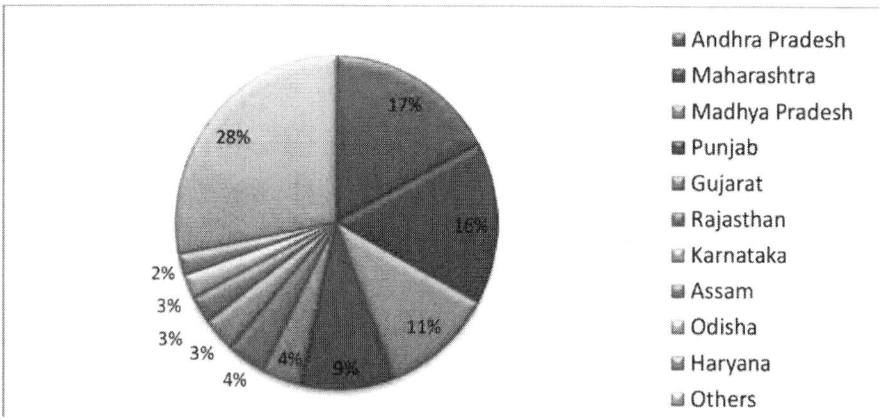

Fig 7.9 Leading citrus producing states in India (2013-14)

In India, Andhra Pradesh top in citrus fruits production (17.16%) followed by Maharashtra (15.79%) and Madhya Pradesh (11.13%) (Fig 7.9).

The data pertaining to the export of citrus fruit (fresh/dried) from India are presented in Table 7.25.

Table 7.25 Export of citrus fruits (fresh/dried) from India

Year	Qty in Tons			Value in Rs Crore		
	Lemons & Limes	Other Citrus	Total	Lemons & Limes	Other Citrus	Total
2006-07	13033	10522	23556	14.04	12.78	26.82
2007-08	16223	9241	25464	15.49	11.12	26.61
2008-09	29494	9896	39390	37.91	10.51	48.42
2009-10	23817	9040	32857	32.12	9.96	42.08
2010-11	28801	6130	34931	34.54	10.44	44.98
2011-12	23510	4820	28329	35.19	9.02	44.21
2012-13	22270	6472	28742	48.57	9.54	58.11
2013-14	19545	5431	24976	52.43	12.51	64.94
2014-15	19033	7074	26107	60.18	13.22	73.4
2015-16	17081	7984	25065	74.07	12.74	86.81
CGR%	0.66	-5.21	-1.31	16.57**	1.08	12.18**
% to total (2015-16)	-	-	-	85.32	14.68	100.00

** Significant at 1% level

This revealed that citrus fruit by quantity did not show any significant growth in export during last decade however by value it grew by 12.18% annually. Citrus export includes lemons and lime and other citrus from which the former contributed 85.32% by value in 2015-16 and reported 16.57% annual growth during last decade.

The detail of major export market for Indian citrus is provided in Table 4.26. UAE alone accounted for 76.42 % share in total citrus export followed by Saudi Arab (12.64%). In case of other citrus, Nepal emerged as a largest market contributing 84.14%.

Table 7.26 Top 10 Export destinations for citrus fruits (fresh/dried) (2015-16)

Rank	Country	Value Rs Crore	% to Total	Rank	Country	Value Rs Crore	% to Total
	Lemons & Limes				**Other Citrus**		
1	U A E	56.60	76.42	1	Nepal	10.72	84.14
2	Saudi Arab	9.36	12.64	2	Bangladesh	0.71	5.54
3	Maldives	3.52	4.76	3	U S A	0.38	2.99
4	Nepal	3.30	4.46	4	Russia	0.33	2.57
5	Kuwait	0.48	0.64	5	Kuwait	0.15	1.16
6	Baharain	0.35	0.48	6	Canada	0.13	1.03
7	Korea	0.12	0.16	7	U K	0.09	0.72
8	Pakistan	0.07	0.10	8	U A E	0.09	0.68
9	Singapore	0.07	0.09	9	Singapore	0.04	0.30
10	Qatar	0.06	0.08	10	Korea	0.04	0.28
Total (Incl. others)		74.07	100.00	Total (Incl. others)		12.74	100.00

Export quality specifications for Citrus fruits are presented in table 7.27. Quality parameters vary from country to country.

Table 7.27 Export specifications for citrus fruits

Mandarin		
Particulars	**Variety**	**Middle East**
Size/Weight/ Colour	Nagpur Mandarin	Colour : Light orange, Weight : 150-175 g Grades : 65-70 mm & 40-45 mm
Packing		65 mm grade – 7 kg 40 mm grade – 2.5 kg
Storage		5-7 ° C
Transport		By Sea
Lime		
Particulars	Variety	Middle East

Mandarin		
Particulars	**Variety**	**Middle East**
Size/Weight/ Colour	Local	20-25 mm, Colour: Green
Packing		5 Kg gunny bags, 20 kg wooden boxes
Storage		5-7 ° C
Transport		By Sea

Melon (fresh)

Watermelon is mainly used as fresh, however it is also used in many versatile preparations such as jelly and sharbat. In India, during the year 2013-14, the area and production of watermelon was 74.64 thousand ha and 1809.83 thousand MT, whereas it was 36.70 thousand ha and 760.81 thousand MT for Muskmelon.

The major producing states of watermelon in India are given in Fig 4.10, which shows that Uttar Pradesh (24.67%), Karnataka (17.53%), Tamil Nadu (12.49%), Odisha (13.42%) and Andhra Pradesh (8.10%) are the five major producing states in the country.

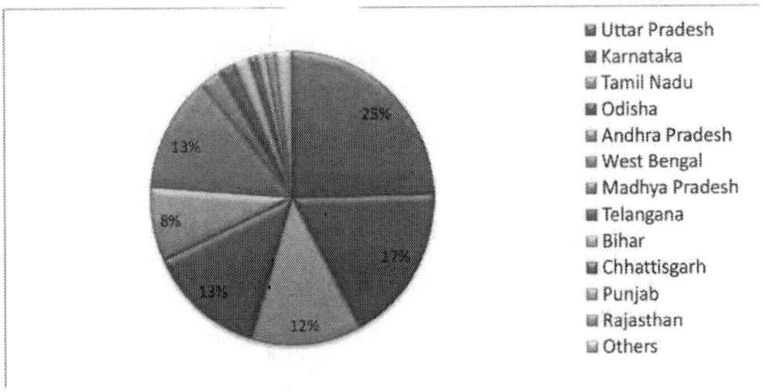

Fig 7.10 Leading watermelon producing states in India (2013-14)

The growth in export of melon (fresh) from India by quantity and value is given in Table 7.28. The group comprises of watermelons as well as other melons. The data showed that the export of water-melon in terms of quantity grew by 10.87% annually and export of other melon also increased even at a higher rate of 17.33% making overall export of Indian melon to rise by 11.77% during last decade. Further, water-melon occupied 67.61% share in total export of melon (fresh) from India by value in 2015-16.

Table 7.28 Export of melon (fresh) from India

Year	Qty in Tons			Value in Rs Crore		
	Water-melons	Other Melons	Total	Water-melons	Other Melons	Total
2006-07	7929	674	8603	6.67	0.83	7.5
2007-08	6619	845	7464	5.08	0.72	5.8
2008-09	9774	1184	10958	9.03	1.45	10.48
2009-10	11855	1721	13575	14.45	2.21	16.66
2010-11	11289	1210	12500	14.49	1.95	16.44
2011-12	17760	1877	19637	21.67	3.27	24.94
2012-13	18724	1644	20369	22.42	2.66	25.08
2013-14	14163	1662	15824	29.16	5.65	34.81
2014-15	15234	2587	17821	36.91	11.7	48.61
2015-16	20196	5422	25618	54.46	26.09	80.55
CGR%	10.87**	17.33**	11.77**	24.46**	35.41**	26.6**
% to total (2015-16)	-	-	-	67.61	32.39	100.00

** Significant at 1% level

Major export destinations for Indian melon (fresh) are depicted in Table 7.29. UAE emerged as the largest export market for both Indian water-melon and Indian other melon contributing 73.57 and 81.41% share in total export in total export value of melon, respectively.

Table 7.29 Top 10 export destinations for melon (fresh) (2015-16)

Rank	Country	Value Rs Crore	% to Total	Rank	Country	Value Rs Crore	% to Total
	Watermelons				**Other Melons**		
1	U A E	40.06	73.57	1	U A E	21.24	81.41
2	Maldives	3.57	6.55	2	Qatar	1.48	5.68
3	Nepal	3.32	6.09	3	Netherland	0.87	3.33
4	Netherland	2.98	5.46	4	Bahrain	0.57	2.19
5	Qatar	1.81	3.32	5	Kuwait	0.54	2.07
6	Germany	1.09	2.01	6	Oman	0.48	1.83
7	Bahrain	0.77	1.42	7	Saudi Arabia	0.29	1.11
8	Oman	0.26	0.48	8	Germany	0.23	0.89
9	U K	0.24	0.45	9	Canada	0.10	0.38
10	Canada	0.18	0.33	10	U K	0.09	0.33
Total (Incl. others)		54.46	100.00	Total (Incl. others)		26.09	100.00

Papaya

Papaya is a tropical fruit having commercial importance because of its high nutritive and medicinal value. During the year 2013-14, area and production under papaya was 133.4 thousand ha and 5639.3 thousand MT (Table 7.30). Production under Papaya crop became almost double during the last decade. In India, papaya is mostly cultivated in the states of Andhra Pradesh, Gujarat,

Karnataka,

Madhya Pradesh and West Bengal. Top producing states of papaya are: Andhra Pradesh (27.39%), Gujarat (21.02%), Karnataka (8.43%), Madhya Pradesh (7.69%) and West Bengal

(5.94%) (Figure 7.11).

Table 7.30 Area, production and productivity of papaya in India

Year	Area ('000 ha)	Production ('000 MT)	Productivity (MT/ ha)
2004-05	73.8	2535.1	34.4
2005-06	67.8	2139.3	31.6
2006-07	72	2482	34.5
2007-08	83	2909	35
2008-09	98	3629	37
2009-10	95.7	3913.5	40.9
2010-11	106	4196	39.6
2011-12	117.4	4457.1	38
2012-13	132.2	5381.7	40.7
2013-14	133.4	5639.3	42.3

Source: NHB

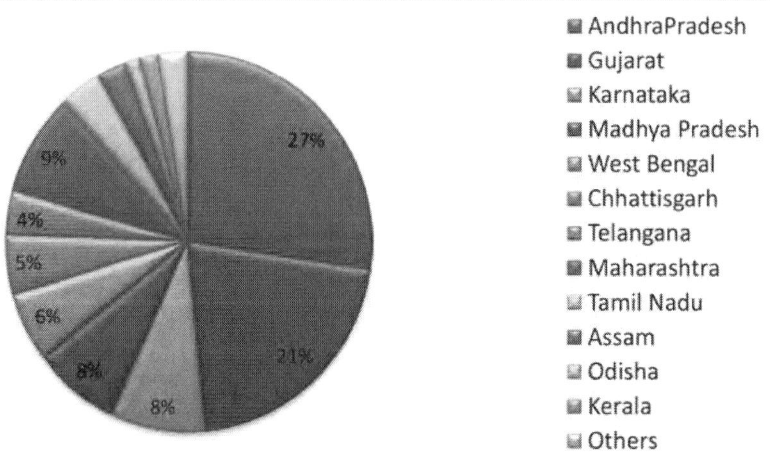

Fig. 7.11 Leading papaya producing states in India (2013-14)

Export of papaya (fresh/dried) from India is presented in Table 7.31. It was found that though the export of papaya by quantity did not show any significant rise during the study period, but by value it increased from Rs 11.60 crore in 2006-07 to Rs 56.78 crore in 2015-16 with a rise of 17.19 % per annum.

Table 4.31 Export of papaya (fresh/dried) from India

Year	Qty in Tons	Value in Rs Crore
2006-07	10344	11.60
2007-08	10880	11.23
2008-09	13834	18.10
2009-10	17925	17.49
2010-11	17176	19.31
2011-12	18672	24.75
2012-13	16491	33.29
2013-14	9922	35.11
2014-15	11485	38.26
2015-16	13181	56.78
CGR%	0.44	17.19**

** Significant at 1% level

Top 10 export destinations for papaya during the year 2015-16 are given in Table 4.32. UAE emerged as the largest export destination for Indian papaya contributing 46.32 % share in total papaya export followed by Saudi Arab (18.80%), Netherland (12.08%), Qatar (6.20%) and Bahrain (5.27%).

Table 4.32 Top 10 Export destinations for papaya (fresh/dried) (2015-16)

Rank	Country	Value Rs Crore	% to Total
1	UAE	26.30	46.32
2	Saudi Arab	10.67	18.80
3	Netherland	6.86	12.08
4	Qatar	3.52	6.20
5	Bahrain	2.99	5.27
6	Kuwait	2.31	4.08
7	Oman	1.42	2.50
8	Nepal	1.25	2.21
9	Germany	0.70	1.23
10	Norway	0.19	0.33
Total (Incl. others)		56.78	100.00

Export quality specifications for Papaya are presented in table 7.33. Quality parameters vary from country to country.

Table 7.33 Export specifications for papaya

Particulars	Variety	Middle East	Europe
Size/Weight/ Colour	Taiwan 785	Pale Green, 1-1.25 kg/fruit	---
	Taiwan 786	Pale Green, 1-1.25 kg/fruit	---
	Solo	----	400-500 g/fruit
Packing		8 kg	5 kg
Storage Temp.		10-13 °C	10-13 °C
Transport		By Sea	By Sea

Guava

It is the fifth most important fruit after mango, banana, apple and citrus in India. Guava fruits are used for both, fresh consumption as well as for processing. Guava is exported in fresh/dried form as well as in form of Jams/ Jellies/ Marmalades. During the year 2013-14, area and production was 268.2 thousand ha and 3667.9 thousand MT in the country. In the recent decade, production became almost double in this crop.

Table 7.34 Area, production and productivity of guava in India

Year	Area ('000 ha)	Production ('000 MT)	Productivity (MT/ ha)
2004-05	161	1682.8	10.5
2005-06	166.5	1736.6	10.4
2006-07	176	1831	10.4
2007-08	179	1981	11.1
2008-09	204	2270	11.1

Year	Area ('000 ha)	Production ('000 MT)	Productivity (MT/ ha)
2009-10	219.7	2571.5	11.7
2010-11	205	2462	12.0
2011-12	219.9	2510.4	11.4
2012-13	235.6	3198.3	13.6
2013-14	268.2	3667.9	13.7

Source: NHB

The leading producing states of Guava in the country are: Madhya Pradesh (22.93%), Uttar Pradesh (16.50%) and Bihar (10.19%).

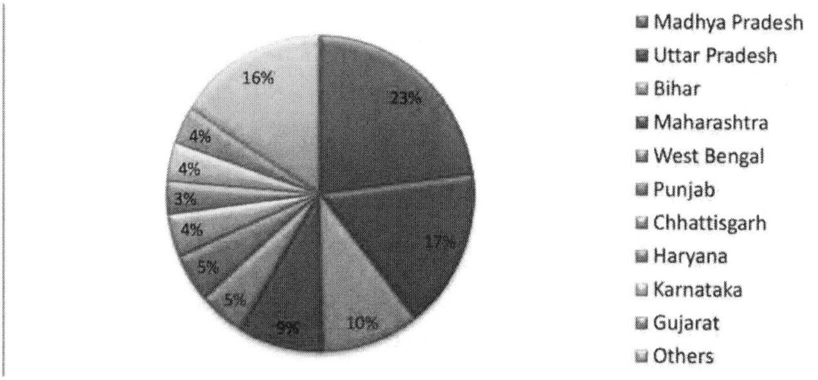

Fig 7.12 Leading guava producing states in India (2013-14)

The growth in export of guava both in terms of quantity and value is presented in Table 4.35. The data revealed that no significant change was observed in terms of quantity but in terms of value it increased from Rs 14.58 crore in 2006-07 to Rs 33.32 crore in 2015-16 indicating 6.23% annual compound growth over the period. The guava is exported mainly in Jams/Jellies/ Marmalades form contributing 78.63% and rest in fresh/dried form.

Table 7.35 Export of guava from India

Year	Qty in Tons			Value in Rs Crore		
	Fresh/ Dried	Jams/ Jellies Marmalades	Total	Fresh/ Dried	Jams/ Jellies/ Marmalades	Total
2006-07	3393	2717	6110	7.73	6.85	14.58
2007-08	2496	5664	8159	4.17	12.4	16.57
2008-09	1691	5471	7163	3.04	15.94	18.98

Year	Qty in Tons			Value in Rs Crore		
	Fresh/ Dried	Jams/ Jellies Marmalades	Total	Fresh/ Dried	Jams/ Jellies/ Marmalades	Total
2009-10	516	5434	5950	1.13	15.37	16.5
2010-11	301	5225	5525	0.62	12.75	13.37
2011-12	1382	4156	5538	3.18	13.71	16.89
2012-13	1180	4570	5749	3.51	16.06	19.57
2013-14	970	3487	4457	4.15	18.1	22.25
2014-15	908	3386	4294	3.72	16.25	19.97
2015-16	1905	5399	7304	7.12	26.2	33.32
CGR%	-6.7	-0.25	-3.25	3.06	8.97**	6.23*
% to total (2015-16)	-	-	-	21.37	78.63	100.00

* Significant at 5% level ** Significant at 1% level

Top 10 export markets for Indian guava in the year 2015-16 are presented in Table 7.36. Saudi Arab emerged as the largest export market for fresh/dried guava with 44.40% share followed by Nepal (21.64%) and UAE (13.70%). Whereas Indonesia, Netherland, UK and Yemen were emerged as top export destination for Jams/ Jellies/ Marmalades accounting 20.34, 16.84, 12.82 and 10.98% share, respectively.

Table 7.36 Top 10 export destinations for guava (2015-16)

Rank	Country	Value Rs Crore	% to Total	Rank	Country	Value Rs Crore	% to Total
	Fresh/ Dried				Jams/ Jellies/ Marmalades		
1	Saudi Arab	3.16	44.40	1	Indonesia	5.33	20.34
2	Nepal	1.54	21.64	2	Netherland	4.41	16.84
3	U A E	0.98	13.70	3	U K	3.36	12.82
4	Oman	0.33	4.70	4	Yemen Rep	2.88	10.98
5	Tanzania	0.24	3.35	5	U A E	1.43	5.47

Rank	Country	Value Rs Crore	% to Total	Rank	Country	Value Rs Crore	% to Total
6	Baharain	0.23	3.29	6	Algeria	1.30	4.97
7	Qatar	0.20	2.78	7	South Africa	1.24	4.75
8	Canada	0.12	1.72	8	Saudi Arab	1.03	3.94
9	Maldives	0.12	1.71	9	Malaysia	0.81	3.08
10	Kuwait	0.08	1.11	10	U S A	0.71	2.72
Total (Incl. others)		7.12	100.00	Total (Incl. others)		26.20	100.00

Sapota

Sapota commonly called chiku is an important fruits of India. Karnataka (26.25% of the total), Gujarat (20.39%), Tamil Nadu (19.97%), Andhra Pradesh (17.15%) and Maharashtra (9.42%) are the leading producing states of sapota (Fig 7.13). In India, during the year 2013-14, the area and production was 177 thousand ha and 1744.3 thousand MT (Table 7.37).

The export of sapota (fresh) from India by quantity and value is given in Table 7.38. The data shows that the export of sapota in terms of quantity declined by 9.99% annually during last decade.

Table 7.37 Area, production and productivity of sapota in India

Year	Area ('000 ha)	Production ('000 MT)	Productivity (MT/ha)
2004-05	134.1	1076.5	8
2005-06	139.7	1114	8
2006-07	149	1216	8.2
2007-08	152	1258	8.3
2008-09	156	1308	8.4
2009-10	158.9	1346.8	8.5
2010-11	160	1424	8.9
2011-12	163.4	1425.8	8.7
2012-13	163.9	1495	9.1
2013-14	177	1744.3	9.9

Source: NHB

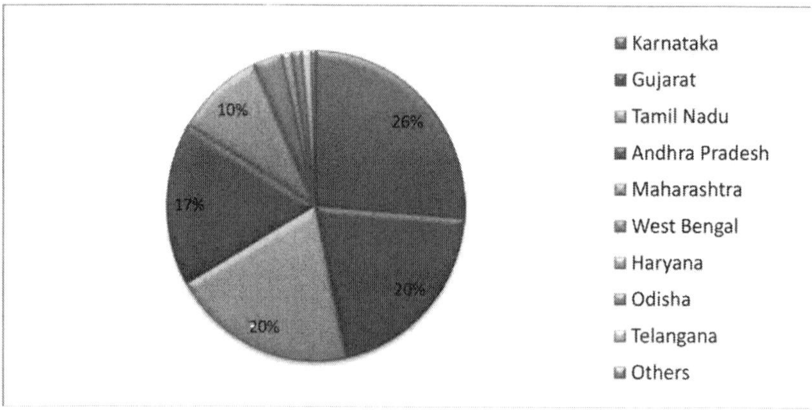

Fig. 7.13 Leading sapota producing states in India (2013-14)

Table 7.39 shows top export destinations for Indian sapota. UAE was the largest export market for Indian sapota which accounted for 49.38% market share followed by Baharain (19.61%) and Quatar (6.29%).

Table 7.38 Export of sapota (fresh) from India

Year	Qty in Tons	Value in Rs Crore
2006-07	2090	7.44
2007-08	2151	5.04
2008-09	4112	6.08

Year	Qty in Tons	Value in Rs Crore
2009-10	4198	5.05
2010-11	2081	3.67
2011-12	2694	4.28
2012-13	2606	6.15
2013-14	1429	7.71
2014-15	1372	8.18
2015-16	973	9.67
CGR%	-9.99*	4.66

* Significant at 5% level

Table 7.39 Top 10 export destinations for sapota (fresh) (2015-16)

Rank	Country	Value Rs Crore	% to Total
1	U A E	4.78	49.38
2	Baharain	1.90	19.61
3	Qatar	0.61	6.29
4	U K	0.56	5.81
5	Saudi Arab	0.44	4.52
6	Canada	0.41	4.26
7	Oman	0.41	4.20
8	Kuwait	0.18	1.90
9	Singapore	0.18	1.84
10	U S A	0.11	1.16
Total (Incl. others)		9.67	100.00

Export quality specifications for Sapota are presented in table 7.40. Quality parameters vary from country to country.

Table 7.40 Export specifications for Sapota

Particulars	Variety	Middle East
Size/Weight/ Colour	Cricket ball, Kalipatti	-
Packing		3 kg
Storage		15-20°C
Transport		By Sea

8

EXPORT OF VEGETABLES FROM INDIA

The export of organic products has picked up substantially in recent years. As the world becomes more health-conscious, such products will grow to be even more popular. Over the years, India has become a significant exporter of organic products due to its agricultural prowess. However, starting an organic food business for export purposes requires meticulous planning and execution.

Due to varied agro-climatic zones, India can grow a wide variety of vegetable crops the year round. With a rise in the standard of living and awareness on benefits of quality food intake, people are now more inclined towards their health and buying healthier vegetables. In today's era of variation, farmers are adapting modern crop yielding improvement methods for higher yield, quality biotic resistance, flexibility to adapt to changing climatic conditions and managing nutrients in the food to make it healthier.

Though historically neglected, the mechanization of agriculture has been considered a major factor in the development and sustainability of vegetable production. Farmers and respective state governments understand the advantage and hence are introducing schemes and programs to help farmers adopt mechanisation at large scale. A lot of work is required in this direction; however, a head start has been made.

Smallholder farmers have the in depthknowledge about their land which is passed on to their next generations, they understand their seeds and crops better as each type of crop requires different planting methods and with the bliss of the latest planting techniques and equipment, they can utilize this knowledge to their full potential and can yield diverse vegetables on their smaller farms. When smallholder farmers grow high value crop, they can yield better returns.

Since vegetables are harvested quickly, 3-4 crops can be cultivated throughout the year to meet the increased demand for vegetables.Therefore, changes in food consumption and demand for particular vegetable can also be met.

India grows the largest number of vegetables from temperate to humid tropics and from sea-level to snowline, Vegetables are excellent source of vitamins, particularly niacin, riboflavin, Thiamin and vitamins A and C. They also supply minerals such as calcium and iron besides proteins and carbohydrates. Vegetables combat under nourishment and are known to be cheapest source of natural protective tools. However per capita consumption in India is only around 130 grams against a minimum of about 300 grams recommended by dietitians.

Most of the vegetables, being short duration crops, fit very well in the intensive cropping system and are capable of giving very high yields and very high economic returns to the growers besides providing better health standards to the people.

India is the world's largest producer of vegetable next only to China with an annual production estimated around 50.99 million tonnes from 4.5 million hectares. However, vegetables occupy hardly 2 per cent of the total cropped area of the country which is very low in view of the national need. Hence it is necessary to increase the production and productivity of vegetable to meet the demand of growing population to ensure better nutrition by adopting improved technology.

Cole vegetables or Cole Crops- The major cole crops are the cabbage, cauliflower, knolkhol, Brussels sprouts and sprouting broccoli. All these vegetables have been originated from the wild cabbage, 'Cole warts' (Brassica oleracae Var sysvestris) which occurs along the meditarranean coasts. All the cole vegetables require cool season for growth and production of quality product. Among these, cabbage and cauliflower are grown on commercial scale in India. The crops like Knolkhol, Brussels spronts and sprouting Broccoli are grown on a very small scale and their area is scattered around.

AREA, PRODUCTION AND PRODUCTIVITY OF VEGETABLES

Vegetables are important constituents of Indian agriculture and nutritional security due to their short duration, high yield, nutritional richness, economic viability and ability to generate on-farm and off-farm employment. Our country is blessed with diverse agro- climates with distinct seasons, making it possible to grow wide array of vegetables. It ranks second in vegetables production in the world, after China. India produces nearly 11 per cent of the world's vegetables.

As per National Horticulture Database, India is producing 169.48 million metric tonnes of vegetables from 9.54 million hectares during 2014-15 (NHB, 2016). Fruits and vegetables together contribute about 92% of the total horticultural production in the country.

Table 8.1 Area, production and productivity of vegetables in India

Year	Area ('000ha)	Production ('000 MT)	Productivity (MT/ ha)
1991-92	5593	58532	10.0
2001-02	6156	88622	10.7
2002-03	6092	84815	11.9
2003-04	6082	88334	9.9
2004-05	6744	101246	10.1
2005-06	7213	111399	10.4
2006-07	7581	114993	10.7
2007-08	7848	128449	11.2
2008-09	7981	129077	11.2
2009-10	7985	133738	11.3
2010-11	8495	146554	11.7
2011-12	8989	156325	11.4
2012-13	9205	162187	11.6
2013-14	9396	162897	12.3
% share to hor- ticultural crops (2013-14)		38.83	58.73

Source: Department of Agriculture, Cooperation and Farmers Welfare,GoI

The production share of vegetables was 58.73 per cent among all horticultural crops, during the year 2013-14. The production of vegetable crops became more than double in the last recent 14 years from 89 million MT to 162.9 million MT. During the year 2013-14, the area and production under vegetable crops was 9.40 million ha and 162.90 million MT, respectively.

Amongst vegetables, India is the largest producer of ginger and okra and ranks second in production of potato, onion, cauliflower, brinjal, cabbage, etc. The vast production base offers India tremendous opportunities for export.

Potato (25.51%), tomato (11.91%), onion (11.50%) and Brinjal (8.32%) are the four major vegetables contributing 57% of total vegetable production in our country. Other important vegetables are cabbage (5.55%), cauliflower (5.26%), tapioca (5.0%), okra (3.9%) and peas (2.4%).

Table 8.2 Area and production of vegetables in India (2013-14)

Fruits	Area			Production		
	'000 ha	% share	Rank	'000 MT	%	Rank
Potato	1973	21.00	1	41555	25.51	1
Tomato	1204	12.81	2	19402	11.91	2
Onion	882	9.39	3	18736	11.50	3
Brinjal	711	7.57	4	13558	8.32	4
Tapioca	228	2.43	8	8139	5.00	7
Cabbage	400	4.26	7	9039	5.55	5
Cauliflower	434	4.62	6	8573	5.26	6
Okra	533	5.67	5	6346	3.90	8
Peas	434	4.62	6	3869	2.38	9
Sweet potato	106	1.13	9	1088	0.67	10
Others	2492	26.52	10	32591	20.01	11
Total	9396	100.00	-	162897	100	-

Source: NHB

The major producing states in the country are West Bengal, Uttar Pradesh, Bihar, Madhya Pradesh, Gujarat, Maharashtra, Odisha, Tamil Nadu, Andhra Pradesh, Karnataka and others. West Bengal ranks first in terms of area and production under vegetables in India followed by Uttar Pradesh and Bihar. Gujarat State ranks 6th in terms of area (6.20% share to India's area) and 5th in terms of production (7.10% share to India's production), during the year 2013-14.

Table 8.3 Major vegetables producing states in India (2013-14)

States	Area ('000ha)	Production ('000 MT)
West Bengal	1380.3	23045
Uttar Pradesh	859.4	18545
Bihar	809.8	15097.8
Madhya Pradesh	628.7	13019.3
Gujarat	582.3	11571.2
Maharashtra	726	10161.8
Odisha	677.3	9433.7

States	Area ('000ha)	Production ('000 MT)
Tamil Nadu	289.7	8678.8
Andhra Pradesh	439.6	8149.8
Karnataka	418.7	7500.7
Others	2584.2	37693.8

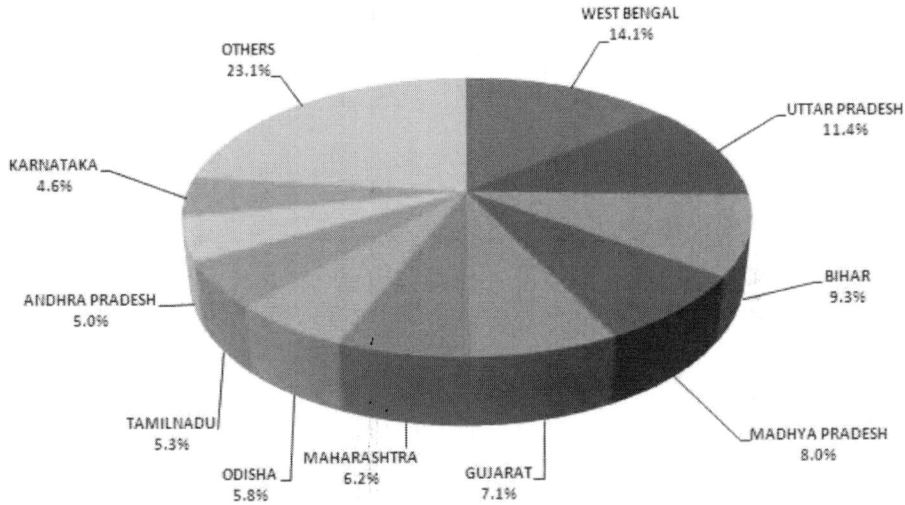

Fig. 8.1 Leading vegetables producing states in India (2013-14)

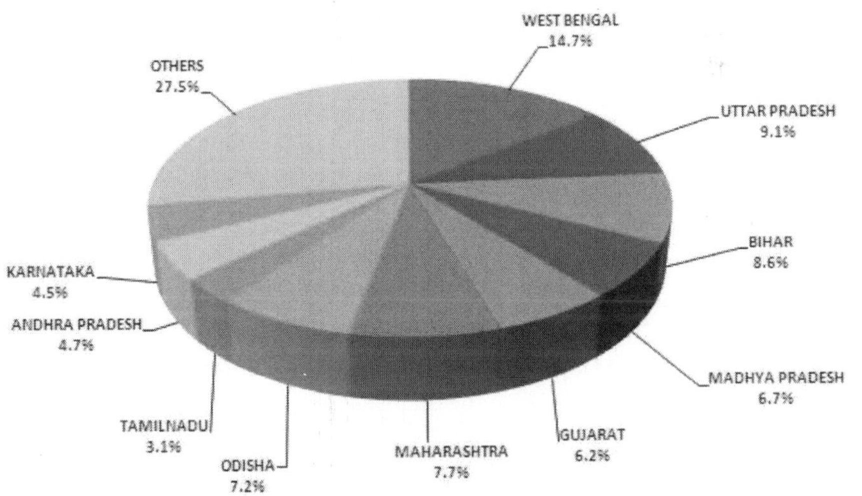

Fig. 8.2 Leading vegetables growing states in India (2013-14)

During the year 2014-15, area and production in Gujarat state, under vegetables was 0.60 million ha and 12.05 MT respectively. Amongst all vegetables,

Potato ranks first in area with 16.23% share and production with 25.71% share.

Table 8.4 Area and production of vegetables in Gujarat (2014-15)

Sr. No.	Vegetables	Area (ha)	% to Total	Prod (MT)	% to Total
1	Potato	98200	16.23	3097391	25.71
2	Cucurbits	81356	13.45	1275094	10.58
3	Brinjal	74014	12.23	1469646	12.20
4	Okra	73843	12.21	857486	7.12
5	Tomato	46228	7.64	1313701	10.90
6	Onion	44500	7.36	1126590	9.35
7	Cluster bean	41752	6.90	421304	3.50
8	Cow Pea	30470	5.04	322084	2.67
9	Cabbage	30029	4.96	654522	5.43
10	Cauliflower	24890	4.11	540954	4.49
11	Other	59684	9.87	970477	8.05
	Total	604966	100.00	12049249	100.00

Source: Agriculture and Cooperation Department, GoG

TOMATO

Tomato is one of the most popular and widely grown vegetables in the world. Tomatoes were introduced in India by English traders of East India Company in 1822. Today, around 10 million 5tonnes of tomatoes were produced in the country with Uttar Pradesh, Karnataka, Maharashtra, Punjab and Haryana being the major producers. The fruits are used as salad, cooked as well as processed for soup, juice, ketchup, puree, paste and powder. Besides table, tomato is one of the important protective foods because of it's special nutritive value. According to the Harvard University study, a diet rich in tomatoes might dramatically decrease the risk of prostrate cancer.

Tomato is warm season crop, and can be grown almost all the year round in milder climate like that of peninsular India. Select medium to heavy well drained soil. For May-June, August-September and December-January planting sow seed in nursery on raised beds 4-6 weeks prior to planting. About

500-600 g seeds/ha are required. Seedling should be transplanted in the field at 60 X 45 or 75 X 60 cm when they are 10-12 cm tall. Apply 75-100 kg n + 100 kg

P2O5 + 50 kg K2O per hectare; in split doses. Half dose of N, full dose of P2O5 and K2O is applied 30 days after transplanting and the remaining half dose of N is applied 30 days after transplanting. Providing staking or support of Banboc sticks to the individual plants or adopt trollies system for training to mato plants. Interculture operations like hoeing weeding irrigation and plant protection measures should be carried out as and when necessary avoid moisture stress during flowering and fruiting stages. Pick tomato at pink stage at an interval 3-5 days average yield ranges from 200 to 300 quintals per hectare.

BRINJAL (SOLANUM MELONGENAL)

Brinjal or egg plant is one of the most common and popular vegetables in India. A number of cultivers if this indigenous vegetable are grown throughout the country. And consumers preference greatly varies, for colour, size and shape. The fruit in India. It is cooked in various ways and Vaishali. Vitally, Puss Purple Round, Puss Kranti, Puss Anmol, Arka Sheetal, Arka Navneet, Punjab-8

The planting is done thrice a year I,e, May-Jane, August-September and December-January. Brinjal requires more warm season. For commercial production of Brinjal well drained fertile soil is necessary. Seedlings are grown at the raised nursery beds. About 400-500 g seeds will be required for planting one hectare area. In a well prepared field 20 to 30 tonnes of well decomposed manure is mixed per hectare and 4-6 week old seedlings are planted at 75 X 60 cm or 75 X 75 cm distance. Apply 60 kg N + 50 kg each of P2O5 and K2O per hectare, of which half the dose of N and full P2O5 and K2O is applied at time of transplanting. Interculture i.e. hoeing, weeding irrigation and plant protection operations are carried out as and when required. The fruits are harvested when they are tender and attractive. The average yield of the fruits varies from 300 to 400 quintal per hectare.

CHILLI (CAPSICUM ANNUUM)

Chilli is one of the most important commercial vegetable crops of India and besides vegetable valued as spices, condiments sauces and pickles. India is largest producer of 6,07,375 tonnes of dry chillies. Andhra Pradesh, Maharashtra and Karnataka are major chilli producing states in India and accounting 75 per cent of the countries chili area.

Chilli can be grown successfully in both cold and warm climatic conditions. Chillies grow on a wide range of soils but for higher yields light loam soil is the best. For rainy season planting seeds are sown in nursery in April-May and for summers planting in January –February. Raised nursery beds of 2m X 1m size with 20 cm are prepared. For planting 1 hectare area 1.5 to 2.o kg seed is required for raising the seedlings.

When seedlings become 4-5 weeks old (10-15 cm heights) These should be transplanted in well prepared mixed with 9-10 tonnes/ha farm yard manure at 60 X 45 cm spacing. Application of 50 kg N + 60-90 kg P2O5 and 50 kg K2O is recommended. About half of the N and full dose of P2O and K2O should be at the time of transplanting and the remaining nitrogen be applied as top dressing after 30-45 days of transplanting.

Interculture including weeding, hoeing irrigation and plant protection should be done as and when required.

Improved Varieties- Pusa Jwala, NP46-A, Pant C-1, G-3, Sindhur, Andhra Jyoti, Agnirekha. Harvesting of chillies depends on the purpose for which they are grown. For vegetables purpose fruits are harvested at green and mature stages, for making powder the fruits are harvested when they are fully ripened, the average yield of chilli as a rainfed crop may be 5- 8 quintal of dry chillies whereas under irrigated conditions 15-25 quintals per hectare, yield can be obtained.

Almost 97.5 % of chilli production is consumed indigenously while only 2.5 per cent is exported. India has exported 34,000 tonnes of chillies valued at Rs 98 crores in 19992-93 while spice oils and oleoresins export was 1,132 tonnes valued at Rs. 5 crores.

OKRA (ABELMUCHUS ESCULENTUS(L-S) MOENCH)

Okra or 'Lady's Finger' is one of the most popular vegetables grown throughout the tropics and warm parts of the temperate zone. The green tender fruits of Okra are rich in Vitamins, proteins, calcium and iron. Mostly it is used in fresh stage but the tender fruits can be preserved by canning in brine, dehydration and freezing. It is also exported as fresh green vegetable to Gulf countries.

It is a warm season crop and thrives best during warm humid seasons, although it grows fairly well in the hot summers.

Okra can be grown in a wide range of soils – sandy loams to clay loams, provided they are well drained and supplied with enough organic matter. Main Okra growing season is Kharif(rainy season) June-Jully. However, it can be grown in September and October and January- February (summer season) where assured irrigation facilities are available.

It requires about 8 to 10 kg/ha of seeds during Kharif and September-October and 18 to 22 kg/ha for the summer season planting. During Kharif season where growth is vigorous, seeds are sown at 60 X 30 cm distance, while in summer the vegetative growth is relatively less and seeds are planted at 45 to 30 cm distance. Seed is sown by dibbling method.

Apply 25 tonnes of farm yard manure per hectare and throughly mix at the time of field preparation. Recommended fertilizer doses are 100 kg N, 50 kg each of P2O5 and K2O per hectare. Full dose of P and K and half dose of N is applied at the time of sowing. The remaining half dose after one month of sowing.

Fields be kept free from weeds, either by hoeing or hand weeding. In summer regular irrigation is given at an interval of 4-8 days depending upon the type of soil; crop growth and climatic conditions. During rainy season, generally no irrigation is given except where there is a long dry spell.

Fruit become roady for first picking after 45 days of sowing. Only tender green fruits be harvested at every 2 or 3 days of interval. It yields about 100 –125 Q/ha in rainy season while the summer season crop yields 40-60 Q/ha.

Improved Varieties- Pusa Sawani, Parbhani Kranti, Arka Anamika, Akra Abhay, Punjab Padmini

Plant Protection-Okra crop is generally attacked by Jassids, aphids, fruit borer, miles and fruit fly insects and diseases like powdery mildew and fusarium wilt. For effective control of these pests and diseases apply 20 kg Furadon per hectare to the furrow before sowing.

Spray the young seedlings with marathion(2 cc per litre of water) a week after the first spray to check the incidence of aphids and jassides. Sprays Nuvacron or Endosulphon on 2 ml per litre of water after 4 weeks and repeated once or twice, to check fruit borer spray Karanthane or wettable sulphur (1 ml per litre of water) to check the powder mildew. Repeat the spray after 8-10 days if necessary.

PRODUCT-WISE EXPORT OF MAJOR VEGETABLES

The product-wise export of major vegetables during the year 2015-16 is presented in Table 8.5 and Fig 8.3. It shows that onion ranked first with Rs 3478.17 crore contributing 51.30% in the total export of vegetables followed by cucumbers & gherkins (14.77%), tomato (6.09%), potato (5.71%), chilli (2.61%), Garlic (2.20%).

Table 8.5 Product-wise export of major vegetables (2015-16)

Rank	Crop	Product	Value in Rs Crore		% share
1	Onion	Fresh/Chilled	2747.41	3478.17	51.30
		Dried/Sliced/Broken/ Powder	730.76		
2	Cucumbers & Gherkin	Fresh/Chilled	2.09	1001.26	14.77
		Provisionally Preserved	365.90		
		Prepared/Preserved	633.27		
3	Tomato	Fresh/Chilled	377.73	412.94	6.09
		Whole Or In Pieces	0.80		
		Prepared/Preserved, / Pieces	1.01		
		Ketchup & Sauces	33.40		
4	Potato	Seeds, Fresh/Chilled	4.80	387.38	5.71
		Other Than Seed, Fresh/ Chilled	320.16		
		Prepared/Preserved, Frozen,	46.85		
		Prepared/Preserved, Not Frozen	15.57		
5	Chilli	Green Chilli Fresh/ Chilled	171.07	177.22	2.61
		Sauce	6.14		
6	Garlic	Fresh/Chilled	73.17	149.45	2.20
		Dehydrated Powder	41.26		
		Dehydrated Flakes	28.05		
		Dried	6.97		
7	Sweet Corn	Frozen	53.48	69.47	1.02
		Prepared/ Preserved, Not Frozen	15.99		
8	Peas	Shelled/Unshelled, Fresh/ Chilled	5.19	22.75	0.34
		Shelled/Unshelled, Frozen	17.56		
	Other	-	-	1081.36	15.95
		Total	-	6780.00	100.00

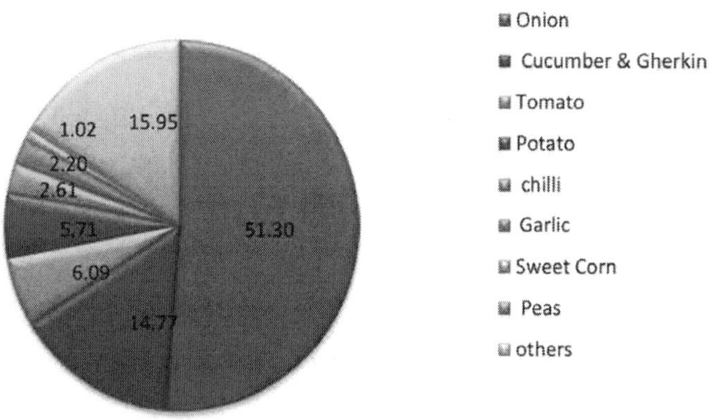

Fig. 8.3 Commodity-wise share (%) in total export of vegetables (2015-16)

Major vegetables Onion

The onion is one of the oldest vegetables known to mankind and is used in an array of recipes and preparations. However, it's rarely eaten on its own and is usually an accompaniment to the main course. India is the second largest onion growing country in the world. Indian onions are famous for their pungency and are available round the year. In India, during the year 2013-14, the area was

1203.6 thousand ha and production was 19.4 million MT, respectivly. Area and production became double in last recent decade in the country. There are certain varieties in yellow onion which are suitable for export in European countries.

Table 8.6 Area, production and productivity of onion in India

Year	Area ('000 ha)	Production ('000 MT)	Productivity (MT/ ha)
2004-05	613.8	7760.6	12.6
2005-06	703.6	9432.5	13.4
2006-07	768	10847	14.1
2007-08	821	13900	16.9
2008-09	834	13565	16.3
2009-10	756.2	12158.8	16.1
2010-11	1064	15118	14.2
2011-12	1087.2	17511.1	16.1
2012-13	1051.5	16813	16
2013-14	1203.6	19401.7	16.1

Source: NHB

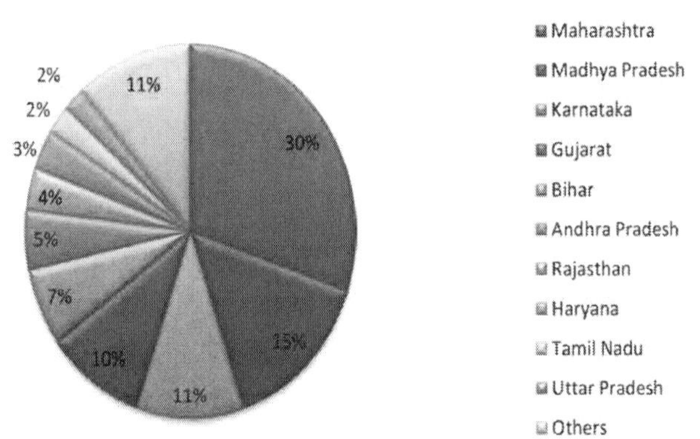

Fig. 8.4 Leading onion producing states in India (2013-14)

The major onion producing states in India are Maharashtra (30.22%), Madhya Pradesh (14.56%), Karnataka (10.64%), Gujarat (9.54%) and Bihar (6.72%). Maharashtra, Madhya Pradesh and Karnataka together contribute more than 50% to the country's total production (2013-14).

Export of onion from India is given in Table 8.7. It was observed that onion export by quantity did not show any significant growth during last decade. This is might be due to frequent and sudden imposed of Minimum Export Price (MEM) by the Government to control export of onion.

The export of onion by its value increased from Rs 1320.66 crore in 2006-07 to Rs 3478.17 crore in 2015-16 showing 10.72% rise per annum during last decade. Further, it was noticed that fresh/chilled onion accounted for major share (78.99%) in total onion export and the rest was in processed i.e. dried/cut/sliced/powder form. Both the categories viz., fresh/chilled and processed by value showed significant growth of 9.42 and 18.18%, respectively during last decade.

Table 8.7 Export of onion from India

Year	Qty in Tons			Value in Rs Crore		
	Fresh/ Chilled	Dried / Cut/ Sliced/ Powder	Total	Fresh/ Chilled	Dried/ Cut/ Sliced/ Powder	Total
2006-07	1378373	24182	1402555	1163.31	157.35	1320.66
2007-08	1008606	59134	1067741	1035.78	230.41	1266.19
2008-09	1670186	27476	1697662	1827.52	143.17	1970.69
2009-10	1664922	28964	1693886	2319.43	238.65	2558.08
2010-11	1182324	38337	1220661	1779.29	287.49	2066.78
2011-12	1309925	56089	1366014	1723.00	388.00	2111.00
2012-13	1666873	51516	1718389	1966.63	458.00	2424.63
2013-14	1482499	40884	1523383	3169.61	509.85	3679.46
2014-15	1238103	52976	1291078	2300.54	679.07	2979.61
2015-16	1201245	53647	1254892	2747.41	730.76	3478.17
CGR%	-0.18	6.36	-0.03	9.42**	18.18**	10.72**
% to total (2015-16)	-	-	-	78.99	21.01	100.00

** Significant at 1% level

Top export destinations for Indian onion are presented in Table 8.8. It was found that in case of fresh/chilled onion export, Bangladesh ranked top with 22.71% share followed by Malaysia (21.34%), Sri Lanka (16.35%) and UAE

(11.91%) whereas in case of processed onion export, Germany ranked top with 13.77% share followed by UK (12.72%) and USA (8.90%).

Table 8.8 Top 10 export destinations for onion (2015-16)

Rank	Country	Value Rs Crore	% to Total	Rank	Country	Value Rs Crore	% to Total
	Fresh/ Chilled				Dried, Whole/ Cut/ Sliced/ Powder		
1	Bangladesh	623.84	22.71	1	Germany	100.63	13.77
2	Malaysia	586.42	21.34	2	U K	92.99	12.72
3	Sri Lanka	449.27	16.35	3	U S A	65.06	8.90
4	U A E	327.28	11.91	4	Russia	52.35	7.16
5	Nepal	173.38	6.31	5	Belgium	47.38	6.48
6	Philippines	75.66	2.75	6	Brazil	38.52	5.27
7	Qatar	69.11	2.52	7	Poland	33.45	4.58
8	Kuwait	66.52	2.42	8	Spain	24.91	3.41
9	Singapore	55.83	2.03	9	South Africa	22.27	3.05
10	Vietnam	52.20	1.90	10	Netherland	17.67	2.42
Total (Incl. other)		2747.41	100.00	Total (Incl. other)		730.76	100.00

Cucumber and Gherkin

The cucumber is a multipurpose vegetable because of wide range of uses from salads to pickles and digestive aids to beauty products. The caloric and nutritional value of cucumber is very low but it is a primary source of vitamins, minerals and fibre for human body. During the year 2013-14, area and production of cucumber was 43.28 thousand ha and 678.15 thousand MT, respectivly in the country.

Andhra Pradesh contributes 23.27% to the India's production followed by Karnataka (15.87%) and Telangana (11.61%). These three states are contributing half of the India's total production. Other producing states are Assam, Jammu & Kashmir, Odisha, Punjab, Madhya Pradesh, Jharkhand, Bihar and others (2013-14).

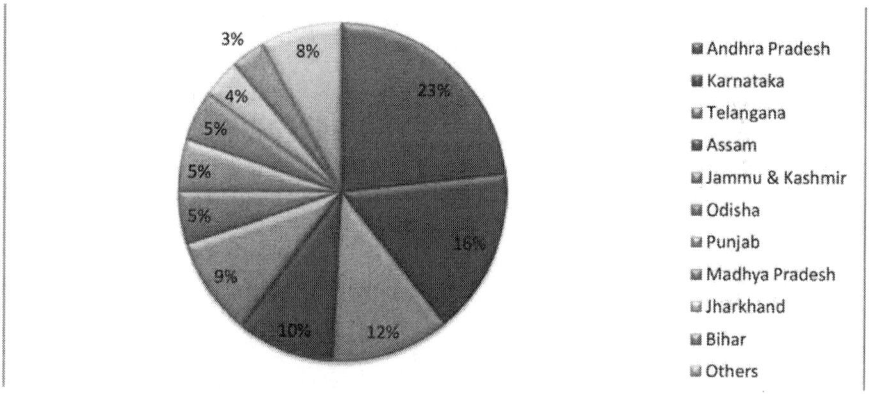

Fig 8.5 Leading cucumber producing states in India (2013-14)

Export of cucumber and gherkin from India is given in Table 8.9.

Table 8.9 Export of cucumber and gherkin from India

Year	Qty in Tons			Value in Rs Crore		
	Provisionally Preserved	Prepared/ Preserved	Total	Provisionally Preserved	Prepared/ Preserved	Total
2006-07	89509	145216	234725	209.04	293.59	502.63
2007-08	60638	127558	188196	145.43	250.62	396.05
2008-09	108081	191910	299992	302.07	487.10	789.17
2009-10	88465	174843	263308	245.16	496.19	741.35
2010-11	59534	149698	209232	139.11	376.15	515.26
2011-12	73709	184894	258603	202.89	542.14	745.03
2012-13	65003	173622	238625	233.24	623.35	856.59
2013-14	51608	167142	218750	241.35	713.86	955.21

Year	Qty in Tons			Value in Rs Crore		
	Provisionally Preserved	**Prepared/ Preserved**	**Total**	**Provisionally Preserved**	**Prepared/ Preserved**	**Total**
2014-15	87432	163751	251183	428.38	774.04	1202.42
2015-16	75412	127514	202926	365.90	633.27	999.17
CGR%	-2.05	0.05	-0.61	7.09	10.77**	9.49**
% to total (2015-16)	-	-	-	36.62	63.38	100.00

** Significant at 1% level

Cucumber and gherkin occupied the second rank in vegetables export from India. It is mainly exported as preserved as well as provisionally preserved. By quantity, none of these categories reported any significant change in last decade. Whereas in terms of value, only preserved category reported significant growth of 10.77 per cent per annum. Overall export of cucumber and gherkin by value reported 9.49% growth per annum. Further, it was noticed that during the year 2015-16, preserved category occupied the larger share (63.38%) in total export of cucumber and gherkin from India as compared to provisionally preserved (36.62%).

Top 10 export destinations for Indian cucumber and gherkin are presented in Table 8.10. It was found that in case of provisionally preserved category, USA ranked top with 23.53% share followed by Belgium (15.42%), Spain (14.29%), France (14.16%) and Russia (13.53%). Whereas in case of preserved category, USA ranked top with 25.04% share followed by Russia (11.02%), Germany (10.43%) and Canada (10.18%).

Table 8.10 Top 10 Export destinations for cucumber and gherkins (2015-16)

Rank	Country	Value Rs Crore	% to Total	Rank	Country	Value Rs Crore	% to Total
	Provisionally Preserved				Prepared/ Preserved		
1	U S A	86.08	23.53	1	U S A	158.60	25.04
2	Belgium	56.41	15.42	2	Russia	69.79	11.02

Rank	Country	Value Rs Crore	% to Total	Rank	Country	Value Rs Crore	% to Total
3	Spain	52.28	14.29	3	Germany	66.05	10.43
4	France	51.82	14.16	4	Canada	64.45	10.18
5	Russia	49.52	13.53	5	Australia	39.04	6.16
6	Iraq	9.34	2.55	6	Belgium	37.66	5.95
7	Germany	5.96	1.63	7	Netherland	28.67	4.53
8	Italy	5.92	1.62	8	U K	26.87	4.24
9	Lithuania	4.92	1.35	9	France	24.87	3.93
10	Australia	4.83	1.32	10	Spain	19.81	3.13
Total (Incl. other)		365.90	100.00	Total (Incl. other)		633.27	100.00

Tomato

Tomato is one of the most important protective food crops of India. Tomato is rich source of vitamin A and C, potassium, minerals and fibres. Tomatoes are used in the preparation of soup, salad, pickles, ketchup, puree, and sauces and also consumed as a vegetable in many other ways. In India, the area, production and productivity was 882 thousand ha, 18.74 million MT and 21.2 MT/ ha in the year 2013-14, respectivly.

Table 8.11 Area, production and productivity of tomato in India

Year	Area ('000 ha)	Production ('000 MT)	Productivity (MT/ha)
2004-05	505.4	8825.4	17.5
2005-06	546.1	9820.4	18
2006-07	596	10055	16.9
2007-08	566	10303	18.2
2008-09	599	11149	18.6
2009-10	634.4	12433.2	19.6
2010-11	865	16826	19.5
2011-12	907.1	18653.3	20.6
2012-13	879.6	18226.6	20.7
2013-14	882	18735.9	21.2

Source: NHB

During the year 2013-14, Andhra Pradesh (17.90%), Karnataka (11.04%), Madhya Pradesh (10.34%), Telangana (7.92%), Odisha (7.40%) and Gujarat (6.72%) are the major tomato producing states in India.

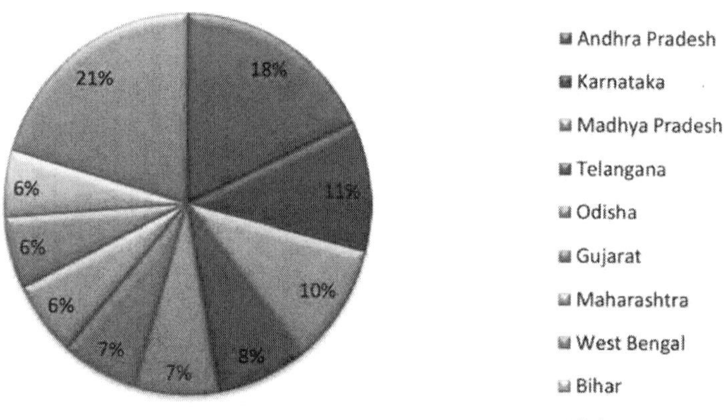

Fig 8.6 Leading tomato producing states in India (2013-14)

Tomato occupied third rank in vegetable exports from India. The data depicted in Table 8.12 shows that export of tomato by quantity as well as by value showed a tremendous growth of 17.14 and 27.13% per annum respectively during last decade. It is mainly exported as fresh/chilled (91.47%) and the rest in processed form. Among the various categories, fresh/chilled and in pieces also showed significant growth of 17.22 and 17.03% per annum by quantity and 27.27 and 20.57% by value, respectively.

Table 8.12 Export of tomato from India

Year	Fresh/ Chilled	In Pieces	Prepared/ Preserved	Ketchup & Sauces	Total
Quantity (in Tons)					
2006-07	33593	135	81	660	34468
2007-08	134845	259	65	1147	136315
2008-09	124617	120	139	1318	126195
2009-10	105862	244	154	1371	107630
2010-11	65872	353	114	8929	75269
2011-12	266986	391	569	2477	270424
2012-13	343692	1272	238	2750	347952
2013-14	426536	616	113	2903	430167
2014-15	217999	1218	775	2729	222721
2015-16	158505	210	190	2907	161812
CGR%	17.22*	17.03*	16.30	14.65	17.14*
Value (in Rs Crore)					
2006-07	34.10	0.39	0.35	2.40	37.24
2007-08	152.91	0.85	0.35	6.54	160.65
2008-09	127.66	0.40	0.77	8.04	136.87
2009-10	103.25	0.71	1.75	8.61	114.32
2010-11	115.10	1.03	0.81	28.08	145.02
2011-12	470.31	1.10	4.75	16.83	492.99
2012-13	533.05	4.08	1.80	18.03	556.96
2013-14	860.91	2.15	1.99	26.87	891.92
2014-15	444.61	6.08	3.47	33.21	487.37
2015-16	377.73	0.80	1.01	33.40	412.94
CGR%	27.27**	20.57*	19.52*	25.95**	27.13**
% to total (2015-16)	91.47	0.19	0.24	8.09	100.00

* Significant at 5% level ** Significant at 1% level

Major export destinations for Indian tomato in the year 2015-16 are given in Table 8.13. It was observed that Pakistan and UAE emerged as major export markets for fresh/ chilled tomato contributing 66.80 and 27.76%, respectively.

For Ketchup & Sauces, USA (29.92%) and Nepal (23.64%) were found major export destinations.

Table 8.13 Top 10 export destinations for tomato (2015-16)

Rank	Country	Value Rs Crore	% to Total	Rank	Country	Value Rs Crore	% to Total
	Fresh/ Chilled				Ketchup/ Sauces		
1	Pakistan	252.33	66.80	1	U S A	9.99	29.92
2	U A E	104.86	27.76	2	Nepal	7.90	23.64
3	Maldives	6.41	1.70	3	Philippines	3.07	9.20
4	Nepal	5.59	1.48	4	U K	3.06	9.17
5	Bangladesh	4.56	1.21	5	Malaysia	1.73	5.18
6	Oman	2.36	0.62	6	Canada	1.14	3.42
7	Saudi Arab	0.99	0.26	7	Israel	1.00	3.00
8	Kuwait	0.32	0.09	8	Bhutan	0.80	2.41
9	Bhutan	0.14	0.04	9	U A E	0.59	1.78
10	Baharain	0.07	0.02	10	Sri Lanka	0.51	1.52
Total (Incl. others) 377.73			100.00	Total (Incl. others) 33.40			100.00

Potato

Potato popularly known as 'The king of vegetables', has emerged as fourth most important food crop in India after rice, wheat and maize. Potato is grown more than 100 countries in world. In India, potato is cultivated in almost all states and under very diverse agro climatic conditions. About 85 per cent of potatoes are cultivated in Indo-Gangetic plains of North India.

Table 8.14 Area, production and productivity of potato in India

Year	Area ('000 ha)	Production ('000 MT)	Productivity (MT/ha)
2004-05	1523.9	28787.7	18.9
2005-06	1569.2	29174.6	18.6
2006-07	1743	28600	16.4
2007-08	1795	34658	19.3
2008-09	1828	34391	18.8
2009-10	1835.3	36577.3	19.9
2010-11	1863	42339	22.7
2011-12	1907	41482.8	21.8
2012-13	1992.2	45343.6	22.8
2013-14	1973.2	41555.4	21.1

Source: NHB

Potato is cultivated in 1.97 million ha area with 41.56 million MT production in the country during the year 2013-14.

The states of Uttar Pradesh, West Bengal, Bihar Gujarat and Madhya Pradesh accounted for more than 80 per cent share in total production. Uttar Pradesh, contributes 31.89% share and West Bengal contributes 22.72% to the total production of the country.

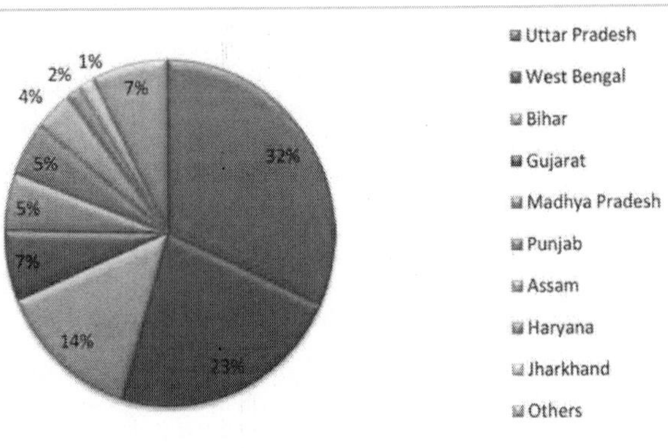

Fig 8.7 Leading potato producing states in India (2014-15)

Potato is exported in various forms viz., fresh/chilled, prepared/preserved (frozen), prepared/preserved (not frozen) and potato seeds. The growth in export of potato from India by quantity and value was worked out and presented in Table 8.15.

Potato emerged as fourth largest exported Indian vegetables. Like tomato, potato also reported a remarkable growth by quantity as well as by value during last decade. Export of potato from India increased from 94240 tons in 2006-07 to 260978 tons in the year 2015- 16 with 13.38% growth per annum. In terms of value, it increased from Rs 71.13 crore to Rs387.38 crore with 24.59% annual growth during the same period.

Amongst various categories, fresh/chilled and prepared/preserved (frozen) showed positive and significant growth by quantity, whereas other two did not show significant change during last decade. By value, all categories except seeds reported positive and significant growth.

Top 10 export destinations for Indian potato (fresh) as well as processed are given in Table 8.16. It shows that Nepal emerged as largest export market for fresh as well as prepared/ preserved (not frozen) potato contributing 52.23% and 82.60% of their respective total exports. For prepared/preserved (frozen) category, Pakistan was the largest export market contributing 54.96%. Egypt emerged as the major market for potato seeds with 49.26% share during 2015-16.

Table 8.15 Export of potato from India

Year	Fresh/ Chilled	Preserved Frozen	Preserved Not Frozen	Potato Seeds	Total
Quantity (in Tons)					
2006-07	89025	758	1298	3159	94240
2007-08	78451	382	1567	3176	83576
2008-09	184961	941	143	11429	197473
2009-10	94088	1671	295	2311	98365
2010-11	185953	135	152	12109	198348
2011-12	193086	1119	336	5669	200209
2012-13	163186	1640	582	1843	167253
2013-14	220926	2256	1446	1734	226362
2014-15	373932	4745	1612	834	381123

Year	Fresh/ Chilled	Preserved Frozen	Preserved Not Frozen	Potato Seeds	Total
2015-16	253135	5862	1033	948	260978
CGR%	13.89**	25.74*	7.61	18.82	13.38**
Value (in Rs Crore)					
2006-07	57.70	4.15	5.98	3.30	71.13
2007-08	41.43	2.97	6.79	4.06	55.25
2008-09	115.04	5.36	0.91	12.03	133.34
2009-10	74.21	9.24	1.50	4.81	89.76
2010-11	155.20	0.95	1.71	10.75	168.61
2011-12	133.45	6.94	5.32	6.02	151.73
2012-13	148.72	10.89	9.98	3.32	172.91
2013-14	251.14	18.43	21.25	3.55	294.37
2014-15	835.06	34.10	21.76	4.67	895.59
2015-16	320.16	46.85	15.57	4.80	387.38
CGR%	25.63**	28.82**	23.84*	-2.08	24.59**
% to total (2015-16)	82.65	12.09	4.02	1.24	100.00

* Significant at 5% level ** Significant at 1% level

Table 8.16 Top 10 export destinations for potato (2015-16)

Rank	Country	Value Rs Crore	% to Total	Rank	Country	Value Rs Crore	% to Total
	Fresh/ Chilled				Preserved Frozen		
1	Nepal	167.23	52.23	1	Pakistan Ir	25.75	54.96
2	Sri Lanka	55.51	17.34	2	South Africa	8.48	18.10
3	Mauritius	21.85	6.83	3	Saudi Arab	2.58	5.51
4	Oman	18.48	5.77	4	Australia	2.56	5.46
5	Maldives	14.00	4.37	5	Canada	1.62	3.46
6	U Arab Emts	11.23	3.51	6	Malaysia	1.13	2.42

7	Kuwait	11.14	3.48	7	China P Rp	1.07	2.27
8	Malaysia	7.22	2.26	8	UAE	0.93	1.99
9	Seychelles	3.18	0.99	9	Singapore	0.66	1.41
10	Hong Kong	2.98	0.93	10	Nepal	0.46	0.98
Total (Incl. other)		320.16	100.00	Total (Incl. other)		46.85	100.00
	Preserved Not Frozen				Potato Seeds		
1	Nepal	12.86	82.60	1	Egypt	2.37	49.26
2	Bangladesh	0.84	5.36	2	Senegal	1.80	37.54
3	Sri Lanka	0.60	3.88	3	Pakistan	0.18	3.68
4	U S A	0.51	3.29	4	Nepal	0.17	3.44
5	Singapore	0.37	2.38	5	Austria	0.16	3.37
6	Malaysia	0.16	1.00	6	Mauritius	0.07	1.53
7	Australia	0.08	0.54	7	Russia	0.04	0.82
8	Thailand	0.05	0.33	8	Kazakhstan	0.02	0.32
9	Seychelles	0.03	0.22	9	Thailand	0.00	0.02
10	U K	0.03	0.20	10	Turkey	0.00	0.00
Total (Incl. others)		15.57	100.00	Total (Incl. others)		4.80	100.00

Green Chilli

Green Chilli is immature fruit that has been harvested before fully ripening. Not as spicy as red chilli peppers, they are available raw, canned or pickled and are a common addition to traditional dishes ranging from soups to snacks. During the year 2014-15, green chilli was cultivated in 170.03 thousand ha area with 1983.45 thousand MT production in the country. Karnataka (32.04%) and Bihar

(21.09%) together contributes more than half of the India's production. Other producing states are Andhra Pradesh, Jharkhand, Haryana, Telangana, Nagaland, Himachal Pradesh and Uttar Pradesh.

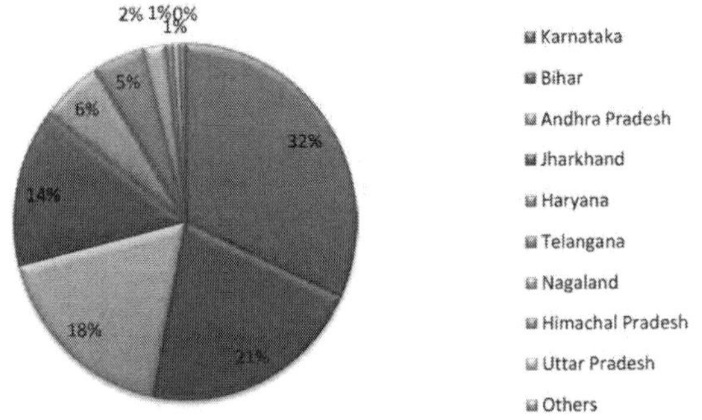

Fig. 8.8 Leading chilli (green) producing states in India (2014-15)

Green chilli is exported as fresh as well as in form of chilli sauce. The growth of chilli export is presented in Table 8.17 and major export destinations for Indian green chilli during 2015-16 are reported in Table 8.18.

Table 8.17 Export of green chilli from India

Year	Qty in Tons			Value in Rs Crore		
	Green Chilli Fresh/ Chilled	Chilli Sauce	Total	Green chilli Fresh/ Chilled	Chilli Sauce	Total
2006-07	18188	126	18314	22.16	0.81	22.97
2007-08	26895	182	27076	43.72	1.03	44.75
2008-09	30198	176	30374	48.32	1.00	49.32
2009-10	30655	72	30726	62.87	0.40	63.27
2010-11	28442	160	28602	61.22	0.94	62.16
2011-12	46703	362	47065	98.24	1.78	100.02
2012-13	35992	369	36361	91.54	2.78	94.32
2013-14	46540	442	46982	140.87	4.00	144.87
2014-15	32138	440	32579	136.14	4.15	140.29
2015-16	39850	575	40425	171.07	6.14	177.22
CGR%	6.94*	18.3**	7.03*	20.18**	25.08**	20.27**
% to total (2015-16)	-	-	-	96.53	3.47	100.00

* Significant at 5% level ** Significant at 1% level

Table 8.18 Top 10 export destinations for green chilli (2015-16)

Rank	Country	Value Rs Crore	% to Total	Rank	Country	Value Rs Crore	% to Total
	Green Chilli Fresh/ Chilled				Chilli Sauce		
1	U A E	104.69	61.20	1	Malaysia	1.65	26.86
2	U K	21.78	12.73	2	Singapore	0.90	14.63
3	Bangladesh	15.09	8.82	3	U S A	0.87	14.18
4	Qatar	13.35	7.80	4	U A E	0.71	11.58
5	Baharain	6.09	3.56	5	U K	0.68	11.12
6	Oman	3.40	1.99	6	Nepal	0.26	4.31
7	Germany	0.92	0.54	7	Vietnam	0.21	3.50
8	Nepal	0.89	0.52	8	Canada	0.21	3.42
9	Australia	0.69	0.40	9	Australia	0.15	2.41
10	U S A	0.65	0.38	10	Sri Lanka	0.08	1.34
Total (Incl. other)		171.07	100.00	Total (Incl. other)		6.14	100.00

The data revealed that chilli export increased by 7.03% annually in terms of quantity and 20.27% in terms of value during the last decade. By quantity, the export of green chilli increased from 18314 tons in 2006-07 to 40425 tons in 2015-16. Similarly, by value, it increased from Rs 22.97 crore in 2006-07 to Rs 177.22 crore in 2015-16. Further, it was found that 96.53% of total green chilli export was in fresh/chilled form.

UAE was the major importer of Indian chilli (fresh/chilled) which accounted for 61.20% share followed by UK (12.73%) and Bangladesh (8.82%). For chilli sauce, Malaysia ranked first with 26.86% share followed by Singapore (14.63%) and USA (14.18%).

Garlic

Garlic is one of the important bulb crops. It is used as a spice or condiment. It has been considered as a rich source of carbohydrates, proteins and phosphorus. It contains volatile oil. Garlic helps in digestion. It reduces cholesterol in human blood. In India, during the year 2014- 15, area and production was 230.6 thousand ha and 1251.9 thousand MT respectively. Madhya Pradesh, Gujarat, Rajasthan, Uttar Pradesh and Assam are the major producing states in India.

Madhya Pradesh contributes 21.57% share to the total country's production followed by Gujarat (19.97%) and Rajasthan (17.45%).

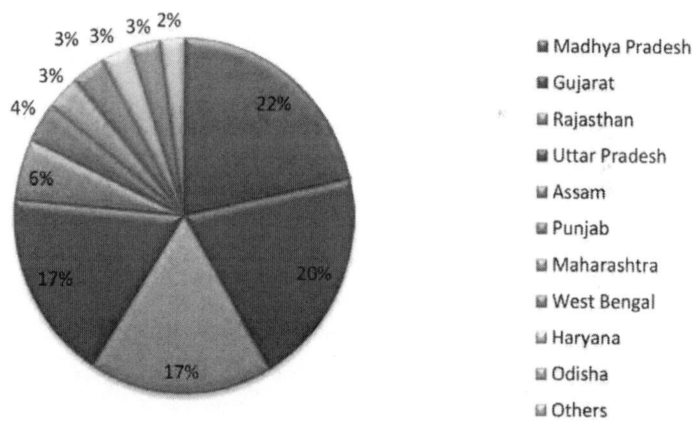

Fig. 8.9 Leading garlic producing states in India (2013-14)

Export of garlic from India as fresh/chilled as well as in proceeded from is reported in Table 8.19.

By quantity, fresh/chilled garlic did not reported significant growth but its value added products viz., dehydrated powder, dehydrated flakes and dried garlic reported significant growth by 27.51, 30.14 and 26.01% per annum, respectively during last decade. By value, overall export of garlic increased from Rs 25.42

crore in 2006-07 to Rs 149.45 crore in 2015- 16 with 30.30% rise per annum. Out of various forms, fresh/chilled contributed 48.96% share in total garlic export in the year 2015-16 followed by dehydrated powder (27.61%), dehydrated flakes (18.77%) and dried (4.66%).

Top 10 Export Destinations for Indian garlic for the year 2015-16 is presented in Table 8.20. It shows that Malaysia emerged as top export destination for fresh/ chilled garlic. Similarly, Thailand for Dehydrated Powder, Vietnam for Dehydrated Flakes and Mexico for dried garlic ranked top in export destinations.

Table 8.19 Export of garlic from India

Year	Fresh/ Chilled	Dehydrated Powder	Dehydrated Flakes	Dried	Total
Quantity (in tons)					
2006-07	12211	779	187	126	13303
2007-08	183	502	252	164	1100
2008-09	859	583	158	70	1670
2009-10	17781	646	426	81	18934
2010-11	21348	1012	1093	333	23787
2011-12	2677	630	629	92	4028
2012-13	36227	3969	1297	393	41886
2013-14	34020	4109	1885	657	40671
2014-15	16079	3689	1187	684	21639
2015-16	15500	4976	2601	955	24032
CGR% 31.48		27.51**	30.14**	26.01** 25.91	
Value (in Rs Crore)					
2006-07	19.31	4.37	1.08	0.66	25.42
2007-08	0.42	2.47	1.88	1.27	6.04
2008-09	2.10	3.01	0.79	0.33	6.23
2009-10	44.52	2.98	2.00	0.57	50.07
2010-11	80.06	5.46	6.97	1.68	94.17
2011-12	9.97	4.52	4.96	0.58	20.03

Year	Fresh/ Chilled	Dehydrated Powder	Dehydrated Flakes	Dried	Total
2012-13	75.59	27.71	7.79	2.94	114.03
2013-14	87.02	26.29	11.74	4.59	129.64
2014-15	42.72	23.36	8.53	3.93	78.54
2015-16	73.17	41.26	28.05	6.97	149.45
CGR%	37.86*	32.29**	34.62**	27.97**	30.30*
% to total (2015-16)	48.96	27.61	18.77	4.66	100.00

* Significant at 5% level ** Significant at 1% level

Table 8.20 Top 10 export destinations for garlic (2015-16)

Rank	Country	Value Rs Crore	% to Total	Rank	Country	Value Rs Crore	% to Total
	Fresh/ Chilled				Dehydrated Powder		
1	Malaysia	22.13	30.24	1	Thailand	6.74	16.33
2	Pakistan	17.64	24.10	2	South Africa	5.17	12.52
3	Sri Lanka	9.40	12.85	3	Spain	3.82	9.25
4	Thailand	7.55	10.32	4	Vietnam	3.26	7.89
5	Vietnam	5.53	7.56	5	Brazil	2.80	6.78
6	Indonesia	4.91	6.70	6	Bangladesh	1.56	3.77
7	Philippines	2.15	2.93	7	U K	1.45	3.51
8	Nepal	1.05	1.44	8	U A E	1.44	3.48
9	Sudan	0.91	1.24	9	Egypt	1.43	3.47
10	U S A	0.56	0.76	10	Turkey	1.35	3.28
Total (Incl. other)		73.17	100.00	Total (Incl. other)		41.26	100.00
	Dehydrated Flakes				Dried		
1	Vietnam	8.11	28.91	1	Mexico	1.73	24.77
2	Thailand	5.32	18.96	2	U A E	1.71	24.48
3	Nepal	4.00	14.27	3	Kazakhstan	0.57	8.22
4	Pakistan	2.33	8.30	4	Nepal	0.37	5.36

Rank	Country	Value Rs Crore	% to Total	Rank	Country	Value Rs Crore	% to Total
5	Mexico	1.58	5.65	5	Poland	0.37	5.34
6	Japan	1.28	4.57	6	China	0.26	3.71
7	Argentina	1.01	3.61	7	Turkey	0.21	3.00
8	Spain	0.72	2.58	8	Serbia	0.20	2.91
9	Nigeria	0.59	2.12	9	Ukraine	0.19	2.74
10	Hong Kong	0.57	2.02	10	Netherland	0.19	2.74
Total (Incl. others)		28.05	100.00	Total (Incl. others)		6.97	100.00

Sweet Corn

Sweet corn is a variety of maize with high sugar content. It is the result of a naturally occurring recessive mutation in the genes which control conversion of sugar to starch inside the endosperm of the corn kernel. Sweet corn is picked when immature (milk stage) and prepared and eaten as a vegetable, rather than a grain. It must be eaten fresh, canned, or frozen, before the kernels become tough and starchy.

Export of sweet corn from India is presented in Table 8.21 and the major export destinations are given in Table 8.22.

Table 8.21 Export of sweet corn from India

Year	Qty in Tons			Value in Rs Crore		
	Frozen	Not Frozen	Total	Frozen	Not Frozen	Total
2006-07	123	32	154	0.34	0.16	0.50
2007-08	1290	28	1318	4.00	0.19	4.19
2008-09	1033	119	1152	2.95	0.49	3.44
2009-10	2279	217	2497	8.11	0.96	9.07
2010-11	3038	38	3076	11.62	0.46	12.08
2011-12	5431	192	5622	19.56	2.38	21.94
2012-13	8341	80	8421	26.73	1.42	28.15
2013-14	10772	24	10796	38.98	0.67	39.65
2014-15	8861	556	9417	56.59	6.22	62.81
2015-16	9965	1446	11411	53.48	15.99	69.47
CGR%	41.98**	27.88	41.18**	49.14**	42.57**	48.23**
% to total (2015-16)	-	-	-	76.98	23.02	100.00

* Significant at 5% level ** Significant at 1% level

It shows that export of sweet corn significantly increased by quantity as well as by value during last decade. By quantity it's increased from 154 tons in 2006-07 to 11411 tons in 2015-16 showing 41.18% annual growth. Similarly, by value also it increased from Rs 0.50 crore in 2006-07 to Rs 69.47 crore in 2015-16 indicating 48.23% annual growth. Further, it was observed that sweet corn is mainly exported in frozen form (76.98%) and the rest in non-frozen form (23.02%).

Russia ranked first with 29.44 % share in export of frozen sweet corn followed by Saudi Arab (27.24%) and UK (19.42%). In case of non-frozen category, Russia again emerged as the top export destination with 85.70% share.

Table 8.22 Top 10 export destinations for sweet corn (2015-16)

Rank	Country	Value Rs Crore	% to Total	Rank	Country	Value Rs Crore	% to Total
	Frozen				Not Frozen		
1	Russia	15.75	29.44	1	Russia	13.70	85.70
2	Saudi Arab	14.57	27.24	2	Saudi Arab	1.01	6.34
3	U K	10.39	19.42	3	U A E	0.67	4.16
4	U A E	6.03	11.27	4	Chile	0.22	1.36
5	Iran	2.42	4.53	5	Malaysia	0.19	1.21
6	Oman	0.96	1.80	6	U K	0.13	0.81
7	U S A	0.49	0.92	7	Norway	0.06	0.38
8	Japan	0.45	0.84	8	Maldives	0.00	0.02
9	Pakistan	0.39	0.73	9	Guyana	0.00	0.01
10	Baharain	0.34	0.63	10	New Zealand	0.00	0.01
Total (Incl. others)		53.48	100.00	Total (Incl. others)		15.99	100.00

Pea

Pea is used as a vegetable or in soup, canned frozen or dehydrate. It is cooked as a vegetable along or with potatoes. During the year 2013-14, the area, production and productivity under country was 433.6 thousand ha, 3868.6 thousand MT and 8.9 MT/ha respectively.

Production of peas became almost double during the last decade in India. Uttar Pradesh, Madhya Pradesh, Jharkhand, Himachal Pradesh and Punjab are the major pea producing states in India. Uttar Pradesh alone contributes 50% of the production of entire country (2014-15).

Export of pea from India is presented in Table 8.24. Pea is exported in fresh/ chilled form as well as in frozen form. The data revealed that the export of pea in frozen category reported significant growth both by quantity (18.75% per annum) and by value (27.61%) during last decade. Whereas, fresh/ chilled pea by quantity did not show any significant rise but by value it reported 19.96% rise per annum.

Overall export of pea reported a significant growth by quantity (18.12%) and by value (25.14%). Further, it was found that frozen category accounted for higher share (77.18%) in total export of peas as compared to fresh/chilled (22.82%) during 2015-16.

Table 8.23 Area, production and productivity of pea in India

Year	Area ('000 ha)	Production ('000 MT)	Productivity (MT/ha)
2004-05	276.3	1944.8	7
2005-06	286.1	2270	7.9
2006-07	297	2402	8.1
2007-08	313	2491	8
2008-09	348	2916	8.4
2009-10	364.9	3029.4	8.3
2010-11	370	3517	9.5
2011-12	408.2	3744.8	9.2
2012-13	420.9	4006.2	9.5
2013-14	433.6	3868.6	8.9

Source: www.nhb.gov.in

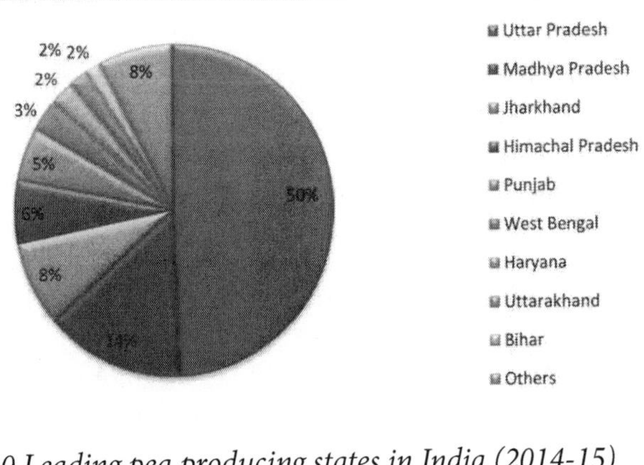

Fig. 8.10 Leading pea producing states in India (2014-15)

The top 10 export destinations for Indian pea are given in Table 8.25. In fresh/ chilled category, UK ranked first with 42.79% share followed by Nepal (29.41%). Whereas in frozen category USA emerged as the top export destination with 32.65% share followed by Saudi Arab (22.80%) and Nepal (13.65%).

Table 8.24 Export of pea from India

Year	Qty in Tons			Value in Rs Crore		
	Fresh/ Chilled	Frozen	Total	Fresh/ Chilled	Frozen	Total
2006-07	1184	949	2133	2.73	2.87	5.60
2007-08	815	1871	2685	2.71	5.56	8.27
2008-09	2519	3526	6046	7.95	12.14	20.09
2009-10	1271	2257	3527	3.54	8.13	11.67
2010-11	1034	4062	5096	2.64	14.28	16.92
2011-12	1981	8565	10546	3.53	32.99	36.52
2012-13	3146	4758	7904	7.65	21.31	28.96
2013-14	8483	11487	19969	20.12	63.77	83.89
2014-15	19366	11758	31124	46.56	65.18	111.74
2015-16	560	2630	3190	5.19	17.56	22.75
CGR%	15.08	18.75*	18.12*	19.96*	27.61**	25.14**
% to total (2015-16)	-	-	-	22.82	77.18	100.01

* Significant at 5% level ** Significant at 1% level

Table 8.25 Top 10 export destinations for pea (2015-16)

Rank	Country	Value Rs Crore	% to Total	Rank	Country	Value Rs Crore	% to Total
	Fresh/ Chilled				Frozen		
1	U K	2.22	42.79	1	U S A	5.73	32.65
2	Nepal	1.53	29.41	2	Saudi Arab	4.00	22.80
3	Saudi Arab	0.42	8.11	3	Nepal	2.40	13.65
4	Australia	0.39	7.48	4	U Arab Emts	1.81	10.31

Rank	Country	Value Rs Crore	% to Total	Rank	Country	Value Rs Crore	% to Total
5	U Arab Emts	0.30	5.74	5	Oman	0.90	5.13
6	Maldives	0.12	2.40	6	Canada	0.71	4.02
7	Ireland	0.12	2.28	7	Kuwait	0.51	2.90
8	Baharain Is	0.03	0.61	8	Baharain Is	0.33	1.90
9	U S A	0.01	0.27	9	Qatar	0.25	1.42
10	Hong Kong	0.01	0.24	10	Singapore	0.25	1.40
Total (Incl. other)		5.19	100.00	Total (Incl. other)		17.56	100.00

EXPORT SPECIFICATIONS

Quality parameters required by major importing countries for vegetables are presented in Table 8.26.

Table 8.26 Export specifications for major vegetables

Sr. No.	Commodity	Parameters for export to Middle East and Gulf countries
1	Okra	3-5 inch length, green tender, packing 5 kg
2	Tondali	1-2 inch length, green tender, packing 5 kg
3	Bottle Gourd	12 inch length, greenish tender, straight, packing 5 kg
4	Pea	5-6 inch length, green tender, straight, packing 5 kg
5	Guar	4-5 inch length, not over matured, packing 5 kg
6	Suran	Cleaned, weighing around 5-10 kg
7	Green Chilli	3-4 Inch length, green, packing 5 kg
8	Drum Stick	24 Inch length, straight, thick, packing 5 kg
	Mode of Transport	By Air or By Sea

Source: MSAMB

9

EMPLOYMENT OF EXPORT ORIENTED HORTICULTURAL CROPS

Agriculture was the mainstay of employment around the world. Since then, the services sector has assumed this mantle and the gap between the two has widened. Although employment growth in agriculture has slowed, the number of workers in this sector reached over one billion, representing 1 in 3 of all workers.

The agricultural sector has some special characteristics that make it more difficult, compared to other sectors, to know precisely how many people it employs. Firstly, in many countries agriculture is still dominated by family farms, where family members provide labour input at different times of the year. Secondly, many farmers and farm workers pursue agriculture as a part-time activity and have other more or less important sources of income. Thirdly, agriculture is characterized by seasonal labour peaks, where large numbers of workers may be hired for relatively short periods.

With 1.3 billion people employed in the sector, agriculture is the second greatest source of employment worldwide after services and it accounts 28% of global employment.

As countries develop, the share of the population working in agriculture is declining. While more than 2/3 of the population in poor countries work in agriculture, less than 5% of the population does in rich countries. It is predominantly the huge productivity increase that makes this reduction in labor possible.

According to the data from 2009, total share of employed in poor countries:

- in sub-Saharan Africa recorded 50% of all employment growth
- in South Asia recorded nearly 33% of all employment growth
- in contrast, in East Asia, Latin America and the Caribbean regions agricultural employment is falling

Agriculture is the most important sector of female employment in many countries, especially in Africa and Asia. At the global level, women do share of 38%, while men about 33%.

Employment in agriculture (% of total employment) (modeled ILO estimate) in India was reported at 41.49 % in 2020, according to the World Bank collection of development indicators, compiled from officially recognized sources. India - Employment in agriculture (% of total employment) - actual values, historical data, forecasts and projections were sourced from the World Bank on July of 2022.

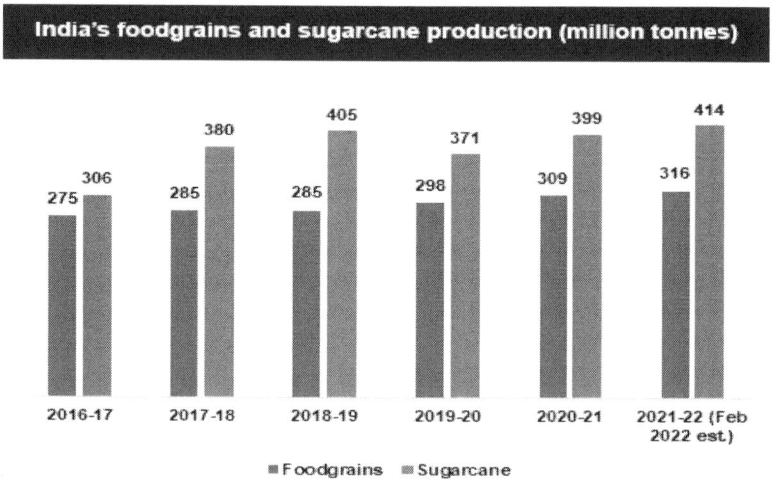

Employment is defined as persons of working age who were engaged in any activity to produce goods or provide services for pay or profit, whether at work during the reference period or not at work due to temporary absence from a job, or to working-time arrangement. The agriculture sector consists of activities in agriculture, hunting, forestry and fishing, in accordance with division 1 (ISIC 2) or categories A-B (ISIC 3) or category A (ISIC 4).

The magnitude of income and employment generation by an enterprise reflects its economic soundness and viability. Income is the ultimate indicator through which impact of any development programme can be assessed. In India, horticultural development has contributed in no small measure towards substantial increase in production and there cannot be two opinions about it. This is substantiated by the fact that increase in productivity of various horticultural crops through various development programmes initiated in this country has brought in its wake financial viability in raising these high value crops for various enterprising producers. However, at this juncture, the relevant question that could strike one is whether various horticultural development programmes in the different parts of the country have been successfully exercise in their influence on various commodity producers in respect of increase in income and employment generation. If yes, to what extend. Another crucial aspect that merits debate and discussion is the determination of magnitude of profit involved in raising these horticultural crops vis-a-vis other field crops, besides evaluation into the pattern of human labour absorption in various operations in the cultivation of these high valued crops. This chapter, therefore, places an onerous task on analyzing these crucial aspects of the selected export oriented horticultural crops grown in the different states are as follows.

ANDHARA PRADESH

Among the three selected horticultural crops, the highest gross and net income are reported by the grape sample cultivators during the reference year compared to the sample households of the other two crops.

Though the cost of maintenance is very high for both the varieties of grapes, the gross income from the crop is also very high leaving handsome profits to the cultivators. Among the different stages of plantation, the gross and net incomes are very high at the stable bearing stage of plantation because of the highest production at that stage of plantation while the lowest incomes are reported at the bearing decreasing stage because of low crop production in both the varieties of grapes.

While very low net incomes are reported from the mango crop during the reference year, negative incomes have been reported by the onion cultivators. Among the various stages of mango plantation, while the highest per hectare net income is reported at the stable bearing stage, lowest net income is reported at bearing increasing stage. While heavy pest attack is reported to be the main reason for low yields from mango crop, unseasonable rains have spoiled the entire onion crop during the harvesting season.

Of the total per household income of Rs.4,77,139, net income from grape crop accounts for 96.08 per cent while income from other sources like wage labour, dairying and other salaried services contributed the remaining 3.92 per cent of the total household income. In the case of mango cultivators, out of the average per household total income of Rs.14,931, the net income from mango crop accounted for only 22.75 per cent while the in come from agriculture (34.68%) and other sources 942.57%) have contributed relatively a higher proportion of income to the sample households.

Of the total household income of Rs.35,983 of the onion sample households, income from agriculture accounted for 80.48% while other sources like agricultural and non- agricultural wage labour have contributed the remaining 29.99%.

As already explained, heavy crop failure due to unseasonal cyclonic rains during the harvesting time of onion crop has placed the sample cultivators at negative incomes they could not recover the paid out costs incurred on the crop during the reference year.

It is found that the quantum of labour employment in horticultural crops is higher than the cereal like paddy. The labour hours in horticultural crops vary 558.35 hours per acre in lady's finger to 1280.57 hours per acre in tomato. However, there is a substantial differences in employing hired and family labour between different size classes. The labour employment in brinjal, potato, cabbage and cauliflower is found to be 1014.93 hours, 914.88 hours, 852.81 hours and 852.54 hours per acre respectively. In case of paddy, it is found that 589.98 hours is needed per acre to cultivate this crop which is substantially lower than the other horticultural crops except lady's finger. Labour employment in orchid per pot is calculated. However, there is another need for labour hours to perform the functions like grading, assembling, packaging etc. specially for horticultural crops which are an essential part in marketing. Thus, in terms of labour employment in all size classes, it is found that horticultural crops are generating more employment than the cereal crops like paddy.

BIHAR

The study reveals that the average labour mandays employed for litchi production in all types of orchard vary from 41 to 49 labour mandays per hectare annually. There has been not any mark difference in per hectare labour employment in marginal, small, medium and large size of orchards. On an average about 46 labour mandays employed per hectare of litchi orchard annually in the study area. However, total labour employed in per hectare litchi marketing and production has been on an average 300 mandays annually. It varies from 289 mandays in marginal orchard to 307 mandays in large orchard. It is worth pointing out that only 15 per cent labourers are employed in litchi production. Moreover, 85 per cent labourers are employed in marketing of litchi.

The study of projection of additional employment generation through litchi cultivation explains that litchi area under Bihar will increase upto 44.43 thousand hectares by the end of 2010 A.D. providing the rate of growth in area during 1981-95 is maintained. Double increase in orchard area would approximately generate additional employment of 88.86 lakh human labour mandays by the end of 2010. It is worthmentioning that the increase in human labour mandays will be mainly in service sector like transportation, packing, storage and retailing, etc. Hence, it may be said that the litchi production will generate employment not only in production process for unskilled labourers but more for the skilled and professionals in the field of marketing. The production of litchi would indirectly generate employment in the core sector of the economy like transport, industry of processing of litchi squash, juice, jam, etc. and persons engaged in production of packing materials.

Income generation through litchi cultivation explains that per hectare income from litchi cultivation was comparatively higher than the cultivation of other field crops in the state of Bihar. At state level the value of total litchi produce was worked out at Rs. 186 crores per annum but all the litchi produced did not find remunerative market. If marketing and processing of litchi is improved the generation of income through litchi cultivation is bound to increase at least two-fold in the state of Bihar.

GUJARAT

Orchard crops are known for their labour intensive and price remunerative characteristics. The employment generated by chiku crop on sample farms was quite significant as compared to field crops.

On the whole, per hectare human labour used for all operations of chiku crop was 399 days. Human labour used for picking operation was as high as 39.2 per cent of total days, which was followed by watching with 24.5 per cent, irrigation with 12.0 per cent and weeding, stocking with 8.8 per cent of total labour days.

The human labour used per hectare on field crops was as low as 162 days. Irrigation, harvesting, threshing and weeding operations accounted for over 80 per cent of total labour days of field crops.

Barring large farm households, per hectare human labour days for chiku crop was positively related with size of land holding groups. The per hectare human labour days for chiku crop was maximum at 521 days with marginal farm households and minimum at 457 days with medium farm households.

The horticultural crops generated better farm income as compared to field crops due to their price remunerative nature. Chiku crop has been an important horticultural crop, known for its sound and steady production, remunerative prices and consequently for better net farm income.

On an average, per hectare net farm income of sample households from chiku crop was Rs. 24126, while the per hectare net farm income from field crops was Rs.18914.

Across the farm sizes for chiku crop there was no relationship between net farm income and size of land holding groups. The per hectare net farm income was maximum at Rs.26787 with small farm households and minimum at Rs.20220 with marginal farm households.

This shows skill and sound organisational capacities of small farm households to earn maximum per hectare net farm income of Rs.26787 compared to households of other size groups with limited resource endowments. Thus, chiku crop has been more labour intensive and price remunerative horticultural crop as compared to field crops.

HIMACHAL PRADESH

The Shimla district is leading in realizing high amount of annual net income from cultivation of fruit crops on all sizes of farms. The per farm annual net returns from fruit crops varies from Rs.23,346 on marginal farm to Rs.2,76,335 on large farm and average being Rs.61,311. The net income from crops other than fruit crops is observed more in Kullu district than in Mandi district.

The per hectare analysis reveals the many fold returns from fruit crops as compared to other crops in all the selected districts as well as on different categories of farms. The per hectare net returns from fruit crops across the farm size are observed low in Mandi district as compared to Shimla district. The weighted average of net returns per hectare from fruit crops of all the districts varies from Rs.45,400 on marginal farms to Rs.71,422 on small farms and average being Rs.55,153. While from other crops the per hectare returns varies from Rs.1,594 on marginal farms to Rs.2,347 on medium farms and average for all is Rs.1,923.

The present study clearly indicates that per hectare net returns from fruit crops are 29 times more than the other crops. Across the farm size, small farmers (having land holding between 1-2 hectares) are realizing more income from per unit of the land devoted under fruit crops. On the small farms the per hectare net income is observed more (Rs.98,520) in Shimla district and less (Rs.11,060) in Mandi district. In Shimla district the share of fruit crops in total income from crop enterprises is hundred per cent while in Kullu the same is around 96 per cent.

The per hectare incremental income from raising fruit crops are Rs.32,411 in Kullu district and average being Rs. 34,957. The change in incremental income is observed quite significant over 1980-89 when the same was Rs. 6,957 in Shimla, Rs. 5,528 in Kullu and Rs. 9,136 in Mandi districts. Thus, it can be concluded that fruit crops give much higher income from the same land as compared to other crops.

In all the production stages the returns from per rupee investment in apple industry in Kullu district are observed better than other two districts under study. The returns from per rupee investment from all the bearing age groups are estimated to be Rs.1.50, Rs.1.58 and Rs.1.66 in Mandi, Shimla and Kullu districts respectively. For the State as a whole the return rate is Rs. 1.57.

The per hectare analysis reveals that small farmers are availing more (301 days) employment as compared to marginal farmers (212 days). The share of hired human labour in production of fruits is observed to be more)94.6%) in Mandi district as compared to 55 per cent in Kullu district and average for all the districts is 75 per cent. On the whole the fruit crops are generating more employment than other crops and all this suggests the high potential of farm employment in cultivation of horticultural crops in the state.

KERALA

On an average per hectare 797 kgs. raw nuts are produced per year. In this the small farmers accounted for 807 kgs. and the big farmers have produced a lowest of 739 kgs. The average gross income for the 40 years of age of plant per hectare works out to Rs.10,90,285 or Rs.27,259 per year and the net income is estimated to Rs.7,99,6\769 or Rs.19,995 per year. The small farmers secured the highest net income than the medium and big farmers because of the relative advantage of family labour work and personal supervision. Any additional labour employed by the big farmers are comparatively costly and they have to depend only on hired labours for all agricultural operations.

All the sample households have raised improved varieties of cashew; many have applied the recommended dosage of fertiliser and irrigation methods; adopted the plant protection measures; marketed their produce in the village itself to the millers; fair knowledge about the price prevailing in the international market and aware of the price variation in different markets.

MAHARASTRA

A comparative economics of growing onion crop vis-a-vis other field crops revealed a much higher net income from onion cultivation compared to the cultivation of other field crops. On an average, the per acre net incremental income accruing from growing onion crop was estimated at Rs. 3043 over that of groundnut and at Rs.4777 over that of soyabean crop. This showed that onion growers improved their net earnings by 64 per cent when they confined themselves to producing onions only instead of groundnut and this improvement was to the tune of 159 per cent vis-a-vis soyabeen crop. In general, the per acre incremental income in growing onion crop over that of bajra, jowar, wheat and gram was estimated at Rs.6821, Rs. 6105, Rs. 6282 and Rs.7561, respectively. Like onion producers, grape orchardists were also found to grow various field crops on their farms. The most profitable field crops grown on the sample farms of grape orchardists were tomato and groundnut. The profit margin in tomato cultivation surpassed even what accrued from onion cultivation. In general, per acre net incremental income in grape cultivation was found to be to the tune of Rs.43,037 over that of tomato cultivation and Rs.47883 over that of groundnut cultivation. It could be worth mentioning that though the cultivation of same of the field crops was not found to be a lucrative proposition, these crops were still grown on the sample farms mainly because of the farmer's need to meet his family consumption requirements.

An analysis of human labour Utilization pattern in various cultivation and marketing operations of onion crop revealed the extent of total absorption of human labour to be to the tune of 98 man days per acre during the kharif season and during rabi season, it was 108 man days per acre. As the land holding size increased, a steady decline in the per acre human labour absorption could also be observed, especially in the kharif season. Splitting of labour Utilization component of onion crop cultivation into men and women revealed that female labour absorption was much higher than the male labour during various operations such as sowing, manuring, fertilising, interculture, mulching harvesting and grading. On the other hand, maintenance of grape orchards was more labour intensive so much so that per acre human labour absorption for grape orchards was equivalent to 200 man days. Also, in grape cultivation, male labour absorption was much higher compare to female labour absorption and this held true for majority of the operations. It was only in the case of manuring and picking/plucking operation that female labour exceeded male labour on grape orchards.

SIKKIM

Regarding the impact of horticultural crops on income, it is found that the growers in the study area have realised more or less steady annual net income from cultivation of horticultural crops than the cereal crops like paddy, maize or wheat. As for example, the cultivation of paddy, maize and wheat generates a net annual income of Rs.1154.86, Rs.1578.86 and Rs.506.83 per acre, whereas cultivation of potato, cabbage, cauliflower and cardamom generates net annual income of Rs.5106.14, Rs.6601.82, Rs.4179.86, and Rs.2706.99 per acre respectively. It is also found that cultivation of orchid is one of the main sources of income of the sample farms in the study area. The share of orchid to the overall net annual income is found to be 79.06 per cent and in terms of money it stands for Rs.44249.13 per farm. In regards to overall incremental income generated from horticultural crops, it is found that Rs.3748.80 per acre is being generated through cultivation of horticultural crops. it is also found that horticultural crops are generating more wage employment than the other field crops in the study area.

It is found that the quantum of labour employment in horticultural crops is higher than the cereal like paddy. The labour hours in horticultural crops vary from 652.54 hours per acre in cardamam to 914.68 hours per acre in potato. However, there is a substantial differences in employing hired and family labour between different size classes. The labour employment in cabbage and cauliflower is found to be 852.81 hours and 852.54 hours per acre respectively. In case of paddy, it is found that 314.41 hours is needed per acre to cultivate this crop which is substantially lower than other horticultural crops. The labour employment in maize and wheat is found to be 589.98 hours and 408.07 hours per acre respectively. Labour employment in orchid per farm is calculated and it is found to be 314.41 hours per farm. However, there is a need for labour hours to perform some other functions like grading, assembling, packaging etc. specially for horticultural crops which are an essential part in marketing. Thus, in terms of labour employment in all size classes, it is found that horticultural crops are generating more employment than the cereal crops like paddy, maize and wheat.

TRIPURA

It has been identified that pineapple and other temperate and tropical fruits have contributed considerably to the state exchequer. The field study revealed that the component of income of the sample farmers from pineapple and other horticultural crops are 74.60 per cent. Pineapple alone accounts for more than 71 per cent: field crops contributed only 2.54 per cent. The non-agricultural sources like services and professions etc. contributed 22.86 per cent.

The findings revealed that the average annual net return from horticultural crops together varied from Rs.17,870 for the marginal farms to Rs.1,64,273/- for the large farms. The per hectare net return found to have varied from Rs. 15,210/- for the large farms to Rs.31,910/- for the marginal farms with an overall average of Rs.20,334/- for the whole sample. The annual per household net return from pineapple fluctuated from Rs.17,13-/- for the marginal farms to Rs.1,57,195/- for the large farms with an overall average of Rs. 44,092/-. Per farm net income for small farms is worked out at Rs.28,314/- Rs.40,307/- for semi-medium farms and Rs.68,786/- for the medium farms.

UTTAR PRADESH

The category wise employment days utilized in various operations on the sample mango farms in Saharanpur district of U.P. indicates that the total employment days utilized per hectare have been estimated to 190 mandays of which the maximum i.e., 174 days have been estimated on account of hired labour and remaining 16 days on family labour which clearly shows that utilization of family labour was negligible on the farms of mango growers. The category wise analysis indicates that the highest i.e., 286 days were utilized per hectare in the category of decreasing production of bearing stage against the lowest i.e., 36 days in the category of non bearing stage. The overall average output per farm from mango orchards has been accounted to Rs.36308. The value of output received from mangoes per farm has been found to be the highest Rs.55570 in the decreasing production stage of bearing orchards against the lowest i.e., Rs.23995 in the increasing production stage. The size group wise distribution of net income from potato in Farrukhabad district of U.P. indicates that the average net income per hectare from potato has been estimated to Rs.86,707. The net income has been estimated to be highest in case of medium farmers followed by marginal farmers.

WEST-BENGAL

Regarding the impact of horticultural crops on income, it is found that the growers in the study area have realised more or less steady annual net income from cultivation of horticultural crops than the cereal crops like paddy. As for example, the cultivation of paddy generates a net annual income of Rs.1824.24 per acre, whereas cultivation of potato, brinjal, cabbage and cauliflower generates net annual income of Rs.5106.14, Rs.16759.36, Rs.6601.87 and Rs.4179.86 per acre respectively. It is also found that cultivation of orchid is one of the main source of income of the sample farms in the study area. The share of orchid to the overall net annual income is found to be 57.40 per cent and monetarily which stands for Rs.45673.87 per farm. In regards to overall incremental income generated from horticultural crops, it is found that Rs.2820.54 per acre is being generated through cultivation of horticultural crops. It is also found that horticultural crops more specifically the orchids are generating more wage employment than the other field crops in the study area.

10

PROBLEMS OF HORTICULTURAL CROP'S EXPORT

As result of several trade liberalization measures and free market environment, a marked change in the trade of horticultural commodities in the form of discernible strategies for their development and new techniques of trade encompassing horticultural exports have emerged. In order to propel Indian economy to a higher rate of growth, the Ministry of Commerce has rightly placed more emphasis on increasing India's agricultural exports through increased fruits and vegetable exports in general and floricultural exports is particular (Anonymous, 1994). However, the crucial question of the hour that may merit attention and fuller discussion is the future prospect of India in horticultural exports. Although India has shown dynamic growth in her exports for most the agricultural commodities in the post liberalization period, to increase the pace of export growth further in the future could be a tough task since domestic requirement of most of the agricultural commodities are also growing very fast due to rise in human population and increase in income levels. As regards agricultural commodities, it is worth mentioning that there are two important aspects which would largely determine India's potentiality to export in the future. These crucial aspects could be traced in the domestic requirement of the various agricultural commodities, particularly horticultural ones, and their expected production levels in the future. In case India generates sufficient production surpluses of specific commodities after meeting the domestic demand, the residual could be made available for exports provided international prices remain favourable.

In this chapter an attempt has been made to study the problems and prospects of export oriented crops in different states of India. Infect, export oriented crops grown on different regions/states have varied type of problems due to variation in agro-climatic conditions in different crop grown in the state. Hence, problems and prospects have been discussed separately for each state.

ANDHARA PRADESH

Fruits and vegetables being highly perishable in nature many problems and constraints have been encountered in their marketing. Some of these problems and constraints are as follows:

Fruits and vegetables are mostly seasonal in nature and available during specific months in a year. The area and production of these perishable commodities are also restricted to specific areas wherever the solid, temperature and other climatic conditions are favourable to them. These perishable commodities arrive for marketing in large quantities during harvest periods creating severe glut in the markets during same year. In other periods, the market arrivals of such commodities are less and far below the internal demand for consumption.

Being highly perishable commodities, fruits and vegetables do not have any regulated marketing in the state except in Hyderabad where a sort of regulated marketing structure exists. In other parts of the state the marketing of fruits and vegetables is largely in the hands of middlemen, namely contractors, wholesale merchants, and retailers. Therefore, the net returns to the grower are low while the profit margin at the tail-end of marketing is high for the middlemen. The inclusion of costs related to transportation, losses due to storage, packing etc. by the middlemen are commonly taken on the higher side pushing up the price of the commodity by the time it reaches the consumer.

The grower has either absolutely no role or a very limited role in collection and distribution of fruits in particular and to some extent, vegetables. The grower is generally satisfied with the early receipts from the crop contractors and does not have any patience or courage to participate in the collection and distribution of these commodities because of highly unpredictable and seasonal price fluctuations.

In the absence of adequate cold storage facilities near the production centre and at the terminal sale point, the grower as well as the merchant is forced to dispose off the commodity at the quickest possible time as the produce gets spoiled in case of prolonged storage.

No standard grading of fruits is being practised so far by any agency except in the case of citrus varieties where the grower is forced to sell out the crop once the crop reaches suitable size and maturity. In the absence of grading, the grower gets lower benefit from the harvest.

Mishandling, inadequate knowledge, lack of awareness for conserving the valuable fruits and vegetables at the right time result in huge wastage at various stages of harvesting, picking, packaging, storage and transpiration. The total estimated wastage during market preparations of fruits and vegetables is to the extent of 25-30 per cent.

Absence of low-cost processing facilities for utilization of the fruit crops and partly damaged fruits in rural area, the loss at early stage is roughly estimated at 5-10 per cent which could be reduced if adequate processing facilities are made available within the reach of the growers.

Most of the fruits and vegetables during their peak harvests enter the markets in large quantities creating a severe glut in the market. In such cases, the harvesting of the crop can neither be prolonged for a few days nor can the produce be stored for a longer duration during the peak season. The grower and the merchants do not play any role to control the glut during such periods of peak arrivals.

Market intelligence reports from different market centres for different fruits and vegetables are presently not made in the country. the available price information for fruits and vegetables is very meagre and is of very limited utility. It is, therefore, essential that accurate data on prices, stocks, arrivals, despatches, etc. of fruits and vegetables are periodically collected from important collection and distribution centres and this data should promptly be made available to producers and traders.

The fruit and vegetable growers are mostly in the unorganised sector and do not have a common platform to voice their problems and grievances and marketing their produce in a systematic and organised way. The state does not have any Fruit and Vegetable Growers' Societies, either at the primary level or at the secondary level, for participation in collection and distribution of commodities. Though there are 3-4 Growers' Co- operatives in the state, most of them are either defunct or not functioning effectively.

BIHAR

In Bihar, the first attempt to export litchi was made by State Trading Corporation during the year 1975 when the corporation tried to export litchi to European countries, but could not yield much success due to lack of infrastructural facilities. Thereafter in the year, 1993 APEDA tried to export litchi through NAFED from Bihar, but the result was not much encouraging as the quality of the fruit was not of the international standard. In the year 1994, APEDA invited Australian consultants to provide export advice for export of litchi from Muzaffarpur. On their recommendation litchi was treated with not benomyl diptreatment before its export, but the same had to be abandoned as the fruits were found "MAGOT" infested. However, in the year 1995, nearly 30 MT of litchi fruit was exported from Muzaffarpur with the help of NAFED to UK, France, Netherland, Mauritious, Singapore and some of the Gulf countries successfully to earn good reputation and appreciation of the quality, colour presentation and packing of the fruit by the foreign buyers. The successful export was possible due to timely spray of phosphomidon arranged by the Bihar State Export Corporation in Collaboration with National Horticulture Board and Plant Protection Department, Bihar. However, this small quantity of litchi could only be exported as against the target of 150 MT of export. The export performance analysis indicated that there is vast potential for export of litchi from the state provided proper infrastructural facilities be developed.

On the basis of the study following policy measures are recommended for improving the production and system of marketing of litchi in the state:

1. As there is immense scope for improving the production and productivity level of this fruit in the state the following measures are suggested:-
   a. Production and distribution of quality planting material through scientifically managed nurseries.
   b. To educate the litchi growers for the maintenance of orchard with respect to manuring and irrigating the orchard periodically as prescribed.
   c. For the scientific guidance, the proper facilities should be provided before the flowering of litchi for proper supervision and timely advice.
   d. Arrangement should be made for supply of necessary chemicals and spraying equipment's as and when required.
   e. New plantation with improved package of practices would be encouraged to replace the old uneconomic orchards.
   f. Arrangements for financial help to the litchi growers at lower rate of interest.
2. For improvement in marketing and export of litchi in the state the following policy measures are suggested:-
   a. The farmers should be encouraged to produce export quality of litchi to get higher price in the world markets.

b. The chain involved in export and marketing of litchi should be minimised.

c. The Government should assure farmers about purchase of all types of fruits. The low quality fruits should be sent for processing of squash, juice, etc.

d. The facilities like pre-cooling and sulphonation chamber, refrigerated van should also be extended to all exporters.

e. The air transport facility should be available in the state capital (patna) to minimise the time taken in export of litchi so that quality of litchi could be maintained.

f. The camp office of APEDA should be in the field (area of orchard) with proper facility so that farmers may get advice of the exports as and when required.

g. The technical expert of APEDA should monitor the post-harvest arrangements at least a week before the export of litchi is commenced.

h. The co-operative institution should be involved in the export of litchi, so that farmers get maximum share in the export of litchi.

GUJARAT

There are good prospects for export of horticultural commodities like mango, banana, onion, potatoes, fresh vegetables from Gujarat. The prospects of export of chiku have been quite limited mainly due to following reasons :

1. Chiku is a highly perishable fruit in nature.
2. The market of chiku is limited upto gulf countries.
3. The expenses of cooling, loading, unloading, transportation to port, insulated vans, customc harges and freight of ships reduced the profit margin in global trade.
4. The cooling treatment, custom clearances and other procedures are too complicated.
5. Generally, foreigners dislike the dark brown colour of chiku pulp.
6. There is an assured market of chiku in northern India.
7. Some times society experiences monopoly of traders.

8. The Amalsad Vibhag Multipurpose Co-operative Society of Valsad Dist., had exported chiku in recent past, but they have stopped it due to low profit, labour intensive work, quality control, lengthy and complicated export processes. Moreover, it was difficult to manage it away from a distance of over 200 kms. from society to Mumbai port.

Nearly 96 per cent of sample orchardists reported to have experienced extreme shortage of farm labour and higher prices of wages. As many as 88 per cent of households had faced erratic and insufficient power supply. Further, they availed power supply during night time.

The shortage of farm yard manure was caused due to dominance of area under orchard crops, which caused shortage of fodder for animals. Hence, there was a shortage of farm yard manure. Nearly 80 per cent of farmers felt the need of having pucca farm roads. Due to clayee, black soil even the kutcha roads remained disrupted during peak seasons.

The village approach roads were in extremely poor condition which caused serious problems of transporting inputs and farm products. Smaller orchards were uneconomic to create owned well or tubewell facility. Hence, farmers had to rely on hired irrigation. Further, it was difficult to watch the scattered orchards during picking period. Due to shortage of labour, picking and grading was done by unskilled labour, which caused problems of improper grading, wastage and lower price.

The chiku markets are located at distance places like Delhi, and hence it had to be transported by trucks, which required packing of hard paper boxes of better quality.

Generally, the product was transported mainly at Delhi market by road in trucks. The road transportation was costlier, and the product was damaged by jerks, moisture and heat during transhipment.

The transportation of horticulture products through railway could be less costly and more convenient. However, the wagons were not available in time. Following are the measures to overcome problems related with crops.

1. Erratic power supply should be regularised with sufficient voltage and that too during day time.
2. Farms should be linked with pucca roads. The village Panchayats, co-operative societies and other local institutions should take initiative and provide some financial aid for construction of pucca road of the farms.

3. The village Panchayats, co-operative societies and other local institutions should take care of regular maintenance of approach roads of villages.
4. Like some industrial goods, horticulture products too should be given priority of allotment of wagons and super fast services.
5. The orchardists should either form co-operative transport societies or the present society should undertake transporting of the product in their own trucks.
6. The co-operative societies should provide timely transport services on farm itself, which will minimise the labour problem and wastage of the product. It will also minimise the problem of glutting of produce in society's yard.
7. Scientific research for better packing should be evolved by National Horticulture Board. Plastic crates or folding boxes should be made available. Research on hard paper boxes which provide protection against moisture should be made.
8. Horticulture Department should encourage farmers to raise demonstration plots.
9. Awareness of farmers regarding picking, grading, packing, marketing intelligence should be increased.
10. Short-term courses of training should be organised by Horticulture Department, co-operative societies, district and taluka panchayats and other voluntary institutions.
11. Soil conservation practices should be adopted by the State Government to check fertility deterioration.
12. At present, the demand of chiku is mainly confined to northern India. If new markets are established in other parts of the country with better infrastructural facilities and sound marketing management, it would help to the better prices as well as to minimise the monopoly of traders.

HARYANA

The global demand for and trade in cut-flowers is increasing at a very fast pace. The area under and production of floriculture products in the developed countries, which are the major consumers of these products, has at best either remained stagnant or declined somewhat. Due perhaps to high cost of labour and infrastructure the prominent developed countries in this trade are concentrating more on technology development rather than on production of flowers.

India has fixed an ambitious export target of floriculture products at Rs.2000 million for the year 2000. Given the pace at which demand for floriculture is rising in the importing countries, achieving this export target does not seem to be an unattainable task. Achieving this challenging target however require developing a high degree of professionalism and improving the capability to compete in the international markets through development of technology and expertise to grow quality flowers at competitive costs. Learning from the experiences gained so far, we would need to think afresh at all stages of project formulation and implementation. A more focused attention on some of the following issues could possibly help in moving towards the goal of realising the targeted export potential and in making Indian floriculture industry a more viable proposition.

Banks and other financial institutions providing finance for floriculture units, during pre- sanctioning appraisal of the projects, have been using certain norms for certain crucial parameters, which as we have reviewed, were much above the actual/realised level. A number of flower exporting units actually suffered losses while some others did not start their operations. The banks have had problems in recovering the money advanced. As a result banks are now apprehensive about advancing such huge sums of money to this sector. If the banks do not come forward to meet the financing needs of the industry, the growth of the industry will be hampered.

The banks therefore need to suitably modify the data base used in their appraisal criterion so as to ensure that unbankable projects are not financed. Given the typical nature of production and marketing process of flowers, any single estimated value for these parameters for projects located in different parts of the country for appraisal purposes is likely to be unrealistic. To take into such likely variations in the values of these parameters it would perhaps be appropriate to start with as realistic numerical values of these parameters as is possible on the basis of available data and then carry out a sensitivity analysis with parametric variations in the values of these parameters - singly as well jointly in various combinations. This will enable determination of a likely range of values over which the project is likely to be viable. Such a procedure will also take care of possible differences in values of these parameters due to differences in location of floriculture units in different parts of the country and will ensure that the unbankable projects are not advanced loans (Indian Banks Association).

The site or location of a floriculture unit is an important contributory factor in determining the cost of a project and its success. While units located in regions/ places having a developed infrastructure need to invest more on climate control, those units located at places which have more favourable climate do not have such developed infrastructure and therefore need greater investment in infrastructure development.

There is therefore a need to carefully evaluate the comparative economics, both from private (entrepreneur's) perspective as well as society's perspective, of locating export oriented floriculture units in different regions. Such an evaluation should also take in to account the comparative economics of developing such units either clustered in a specific geographical region or dispersed in different regions. In the latter case the optimum size of the industry in different regions would also need to be estimated (Directorate of Horticulture, Govt. of India).

Floriculture export being a relatively new enterprise, an organised and efficient marketing intelligence network in this area is wanting and a need for this has often been felt. There is therefore a need to systematically collect and analyse data on such aspects as demand pattern, market arrivals, prevailing prices, price determining factors such as changing consumer's tastes and preferences for different varieties/species of flowers etc. so that entrepreneurs can fetch a better price for their produce and a viable floriculture industry can be developed, (APEDA, NHB, Association of Producers).

Setting up of a green house (GH) is often the single largest item of capital investment for a floriculture unit. Most of the green houses installed in the existing units so far have been imported from foreign countries at exorbitant costs even though indigenous capabilities for building GHs does exist. Much however is not not known about the comparative technical and qualitative features of the two types of green houses. Besides technical assessment, there is a need to evaluate the comparative economics of investing in an imported or an Indian made GH through a careful comparison of the capital and operating cost of the two types of GHs over their life spans. Such an evaluation of comparative economics, must take in to account the differences, if any, in quality and therefore in the price realised of the product produced under the two types of GHs. Such an analysis will enable Indian manufacturers of GHs to carry out the necessary modifications in design/quality of material used in fabrication of GHs. (Floriculture industry and/or government can possibly provide funding for such R & D efforts in this direction).

While a large number of developing countries exporting flowers to EU are not levied any import duties by EU. Indian cut-flower exported to EU attract an import duty of between 15 to 20 per cent. This puts the Indian Exporter at a disadvantage vis-a-vis his competitors. The government, through diplomatic channels, should take up this matter with EU so that Indian cut flower exports are treated at par with exports from other developing countries. Till the time this is accomplished, the government can consider reimbursing, either in full or in part, the duty levied by the EU (APEDA).

The lack of availability of adequate and quality infrastructure for post harvesting care and transportation has been an important contributory factor in realising lower than the anticipated proportion of exports to production as also lower than expected export price realisation. Development of exclusive cold storage at airports, is absolutely necessary (APEDA).

The sudden spurt in export oriented floriculture units in the country in a span of four to five years has created demand for trained manpower to organise and manage cultivation of flowers under protected conditions and undertake their marketing in foreign markets. The training curriculum imparted in our agricultural universities/institutions is however not geared to meet the requirements of this growing industry. There is therefore a need to suitably incorporate in the course curriculum of such institutions, certain courses to meet the specific needs of this industry. Till the time such changes occur, short term training programmes/refresher courses can be organised for existing agricultural graduates to fill the gap (ICAR).

There is a need to diversify our exports to a number of new markets rather than depending mainly on EU countries and U.S.A. for our exports. The successful exports of Indian roses to Japan is an example. A number of countries import flowers from EU to meet their flower requirements. Efforts should be made to systematically analyse the consumption requirement and behaviour of such markets with a view to making direct exports to these markets (APEDA).

Investing in a high cost export oriented enterprise producing a perishable commodity with a very short self life, as in the case of cut-flowers, runs a number of risks most of which may be beyond the control of the entrepreneur. To minimise the risks associated with such an enterprise, on account of marketing exclusively in a foreign market, there is a need to concurrently and systematically develop domestic market as well so that the exporter is assured of a market on which to fall back upon for non-exportable grades of flowers or the excess flower production which somehow can not be profitably exported. This besides meeting the demands of domestic consumers for quality flowers will also enable the producer to cut down on his unit cost of production (APEDA, NHB).

Some of the smaller countries like Thailand, Kenya, Ecuador etc. have a much larger share of international trade in cut-flowers than India. Apart from the fact that these countries have been in this trade for a long time and have established a sort of reputation for themselves and developed contacts in the trade, we need to understand the production and export policy scenarios of such countries and analyse their cost structure. We also need to understand the level of their technology development and nature of technical collaboration they have entered in to with technology supplier so as to understand how they respond to market demand for different varieties and quality of flowers (APEDA could possibly sponsor some such study).

HIMACHAL PRADESH

In Kullu and Mandi districts, the shortage of suitable plant material as well as high cost of the same is reported by more than 50 per cent of growers. High price of fertilisers was also reported by 48,39, and 24 per cent of sample growers respectively in Mandi, Kullu and Shimla districts. More than 30 per cent growers have also reported the scarcity of labour and high rate of wages during peak seasons of important operations of apple cultivation. The maximum numbers of sample orchardists have also shown their serious concern about sudden out break of apple diseases like scab and premature leaf fall etc. They are worried for timely solution from scientists for the eradication of these diseases. Almost all the selected orchardists reported the problem of poor conditions of roads and most of the kuchha roads remain disrupted during peak harvesting season of apple crop. This ultimately hampers the movement of produce to marketing centres and reduction in quality of produce which results in low prices. The high cost of alternate packing material i.e.c.f.b. cartons is also reported by maximum number of orchardists in all the districts. It is interesting to note here that apple packed in cfb cartons are sold at high prices because more quantity of apples is packed in carton as compared to wooden box. This aspects need through investigation to assess the popularity and viability of carton packing in Himachal Pradesh. The high cost and untimely availability of transport facilities were also reported by sample growers.

The maladies of apple scab and recent appearance of pre-mature defoliation of leaves need sincere efforts on the part of horticultural scientists and state horticulture department to equip them selves for control of these diseases on war footing. Sincere efforts are also required to train and create awareness about the latest technical know- how of the production technologies to the farmers.

In addition to production technology the need for creation of awareness about post- harvest management of fruits among growers is quite essential. There is need to train growers about the preparation of fruits for market, market rules and regulations and information about existing market infrastructure available in the state. For reducing spoilage during the transportation of fruits, there is need for development of cool chain system of transport so that produce can be sent in good quality to distant markets.

KERALA

- Cashew has a steady demand and the requirement of the industry is more than 8 lakh metric tonnes a year, whereas the availability from internal sources is less than half of it. The waste and dry lands in Tamil Nadu, Andhra Pradesh, Karnataka, Orissa and Maharastra should be brought under cashew cultivation.
- The existing old unproductive trees have to be replaced by new H.Y.V. crops and modern technology on fertilisation, pruning and grafting and high density planting of cashew trees should be adopted.
- The present Cashew Development Board has to stretch its infrastructure development activities simultaneously to Tamil Nadu. Andhra Pradesh, Karnataka, Orissa and Maharastra.
- Positive financial assistance to the farmers and processing industrial units will increase the rural employment and the growth of the country.
- Kerala has less waste lands when comparing with other southern states. Potentialities to double the cashew production and export earnings are in the reach of the southern states.
- The Government of India should grant plantation status to cashew. Sales tax and land ceiling should be removed.
- Unutilized waste lands under forest control should be distributed to the cashew growers. The Eco-friendly tree will serve as a forest tree as well as an export earning crops.
- Government of Kerala should return the cashew factories retained by them after expiry of the lease period, as this would provide more employment to cashew processing unit.
- Cashew Development Board should devise new norms to have direct linkage of wages to productivity. Cashew apple is highly nutritious and can be utilised for preparation of different fermented and non fermented products namely juice, syrup, jam, candy, pickle, vinegar and liquor varieties like wine, brandy etc..

MAHARASTRA

Two differing scenarios were noticed in terms of export trade of India in grapes and onions over the past decade and a half. While on one hand grape exports from India increased by leaps and bounds during the period between 1979 and 1993, onion exports from India grew very slowly during this period, in fact, during some phases of this overall period the export of this high value crop from India was seen to decline, both in quantity and value terms. Overall, onion export trade of the country could still show some encouraging trend and it was found that in quantity terms, the union exports from India had expanded by 163 per cent between the 1979-1993 period and this expansion in value terms was of the order of 215 per cent. Most of the export expansion for onion was noticed between the early - and the mid-eighties and, thereafter, it had considerably slowed down. In contrast to slow growth of onion export from India, grape exports were found to register sharp increases, particularly after the mid-eighties so much so that during the period between 1985 and 1993 the grape exports from India rose by 525 per cent in quantity terms and 370 per cent in value terms. During 1993-94, UAE, Bangladesh and Malaysia were the major importers of Indian onion, who along with Sri Lanka, Singapore and Saudi Arabia together accounted for 95 per cent of the India's export of onions, both in quantity and value terms. During the same period, the major importers of Indian grape were UAE,UK, Saudi Arabia and Bangladesh.

Maharashtra tops the country in the production growth of many of the horticultural crops, particularly various fruits like grapes, oranges and chewnuts. Notwithstanding such impressive credentials for Maharashtra in horticulture crop production, the share of this state in the nation's total production of banana and onion has often been marked with declining trends which could be construed as depressing because these two commodities have long been at the forefront of India's horticultural exports. Two differing scenarios were perceptible in Maharashtra's trend of grape and onion production. While Maharashtra's grape production recorded impressive and significant growth owing mainly to perceptible rise in productivity and acreage under the crop, the state appeared to lose its share in the nation's total onion production due to losses in productivity. The decline in productivity inspite of reasonable growth in acreage under the crop meant slow growth for the state in onion production. It might appear that the onus of technological efforts have been more favourably inclined and concentrated behind the cultivation of grape rather than onion in the state of Maharashtra. Although both onion and grape turned out to be profitable crops in this state, the element of profit involved in the cultivation of grape was certainly much higher compared to onion. In fact, annual profit from grape cultivation was found to be 3 to 4 times higher than that accruing from onion cultivation. No wonder therefore that more of the farmers were switching over to better profitable grape cultivation. One of the reasons for slow acreage growth under onion compared to grape can be traced in this changing cropping pattern in favour of grape.

One of the important features of grape cultivation in the state of Maharashtra as observed in this study was the practice of pre-harvest contract adopted by majority of the orchardists. Pre-harvest contracting not only facilitated risk-free welcome finance for the orchardists, it also spared them of the botheration of looking after the crop through to the harvest and marketing. However, mention may be made here that this practice of pre-harvest contracting definitely reduced the actual profits accruing to orchardists and, also discouraged them from producing the crop on a large scale. In the ultimate analysis, this may not only affect the economic position of the farmer but also the cultivation of this high value crop. As a matter of fact, a higher production performance can be achieved only by providing a complete package of necessary post- harvest infrastructural facilities. Not much has been achieved in this direction and there is practically a vacuum in this regard. Post-harvest operations such as cooling, grading, packing, transportation, warehousing etc. involve enormous efforts and investments which the Government alone cannot provide. It becomes, therefore, necessary for the farmers to come forward, join hands and collectively share the responsibility. In this respect, the government's endeavour should be limited to equipping the producer with the latest technology, tools equipments and facilities so that they are able to offset and minimise initial post-harvest losses and also slow down the chink of reactions that trigger spoilage.

As regards employment generation through horticultural crops, both onions and grapes were seen to generate sufficient employment opportunities for both female and male workers. However, male workers had better employment opportunities on rape orchards while females were the preferred lot in onion cultivation.

Indian grapes and onions are arguably valued commodities in world trade. The imperative need of the hour is to cut down upon the past-harvest losses in order to meet their increasing demand in the international market. Marketing systems encompassing grapes and onions also need improvement in their efficiency so that the producer has a better stake in the consumer rupee involving these two commodities.

In this changing liberalised economic era, India should give boost to its horticultural production and international trade in horticultural commodities through a concerted drive towards providing proper technological back up, improvements in productivity, efficient marketing, attractive price support and appropriate extension services. The country's international horticultural trade can definitely have a cutting edge over several of the current player nations if our current efforts at giving a major fillip to horticultural production continue unabated with a unflapping zeal.

SIKKIM

The nature and extent of problems/constraints in producing the horticultural crops as perceived by the sample farmers have been thoroughly assessed. There are seven field level problems/constraints according to the perception of the farmers which play as a powerful barrier in production of horticultural crops in general and the orchid in particular. However, the degree of intensity of these constraints vary from farm to farm which is further clearly displayed in the analyses. In marginal farms non availability of skilled labour is the main constraint whereas in small farms, the main constraints is low prices of output. But there is almost close identity in perceiving the constraints in between medium and large farms as projected. It is found that low prices of output is the main constraints in these farms.

In marginal farms, the degree of intensity of constraints according to their importance may be spelt out as lack of working capital, low prices of output, lack of information about improved technologies, non-availability of skilled labour, non-availability of transport and non-availability of proper marketing facilities.

Orchid in particular and the horticultural crops in general being an export potential crops of the state, the prosperity of the growers as well as the overall economy of the state is closely interlined with the development of the production process and marketing methods of these commodities. The result of the present study clearly demonstrates that there is a promising trend towards cultivation of these crops throughout the state as these crops are not only profitable in terms of higher output-input ratio but also generate a higher wage employment for agricultural sector.

However, for adequate growth of this sector vis-a-vis expanding of export of these products attention may be made in developing appropriate organisation. In this juncture a change in the land ceiling act may be amended for allowing the corporate companies to invest more capital on development of large sized captive gardens. There is also an urgent need to adopt effective steps to fulfil the objectives of transfer of technology (TOT) in a proper manner. In this connection, for effective agricultural extension, a strong need is felt for a closer co-ordination and monitoring between the Agricultural Directorate and Horticultural Directorate of the state. Agricultural extension workers can play more effective role in extension and development activities, as well as in transferring the latest technologies at the door step of the growers.

TRIPURA

Most of the pineapple growing areas in the hills are connected by bridle path or kutcha fair weather road. Such roads are not useable for motor transport for carrying of harvest. In the hill areas of the N.E. only one village out of 20 is connected by motorable roads. Obviously such road transport cannot serve the need of the pineapple growers and the farmers are forced to sell his produce in the village to the middlemen at throw away price.

For perishable commodity like pineapple cold storage facility is necessary to maintain its quality in fresh form. Due to lack of such facility the farmer had to sell the product immediately after harvest at whatever price offered by the traders. The middlemen took full advantage of the situation and exploit the growers from their due share of the produce.

There is no regulated market for agricultural produce to cope with the demand for such facilities. The number of unregulated market is substantially large which cheated the farmers by adopting various fraudulent practices. to improve the efficiency of agricultural marketing and to save the farmers from exploitation and malpractice's of middlemen emphasis has been laid on the development of farmers co-operative marketing sociality. Through such societies only the farmers may take the advantage of collective bargaining to obtain a fair price for the produce.

The farmers do not have information on market demand and prevailing prices of pineapple in different markets. Most of the farmers do not have any contact with the central market; as a result they accept whatever price the trader offers to them. So far as fruit crops are concerned media like Radio, Television and Newspaper coverage on market prices is very limited. The interior hill farmers do not have access to such media also and as such the middlemen's offered rate is the prevailing price in that area.

To promote export the following issues must be considered. Emphasis should be given on the development of varieties with higher production potential and export oriented quality and standard. Horticultural research efforts is needed to improve the quality of fruits through adoption of bio-technology. There is need for investment in horticulture sector particularly in pineapple and other export oriented horticultural crops to cope with the demand for fruits in the export market. Improved processing industries are necessary to promote export. Systematic market surveys should be conducted on demand of fruit and fruit products in different countries and some private exporters of horticultural products may be encouraged with incentives to promote export. Cold storage facilities should be created at central market alongwith refrigerated carrier facilities from the farmers field. For export of fresh fruits to the neighbouring countries pre-packaging treatment and use of improved packaging materials should be encouraged, so that, fruit quality is not deteriorated. Air freight subsidy should be given to encourage export of horticultural crops and space for air-carge should be increased for promotion of export of fruits and fruit products from N.E. India. Opening of liberalise boarder trade with bangladesh is necessary as there is considerable demand for Indian fruits like pineapple, jack fruits, citrus and wood apple. Establishment of industry to manufacture yarn from pineapple leaves may be one of the viable proposition which has export potential.

UTTER PRADESH

The problems of plant protection materials were reported to be acute on the farms of almost all the farmers. Operational difficulties were also reported by the majority of farmers of all the bearing stages.

Shortage of skilled labourers on the farms of bearing stages for packing and grading was reported. Prices of packing materials were reported to be much higher. Shortage of gunny bags, wooden boxes and other materials was reported on almost all the farms. Transportation charges were reported to be very high by farmers of all the categories. There was no problem of storage.

Regarding wrong pricing policies it was reported that prices were low and were not announced timely and were not paid timely. Lack of technical know how was there on the farms of all categories.

Regarding measures of improved practices, the best source of extension media was exhibitions/fairs. The services of V.D.C. (Village Development Officers) were reported to be the best and common source to the maximum numbers of mango farmers. The other sources were horticulture officer. plant protection officer, agriculture inspector, agriculture extension officer, Block Development Officer, progressive farmers, friends and relatives.

The problems faced by potato growers were mainly the non availability of electricity and diesels, seeds and fertilizers, plant protection materials and approach roads.

There was shortage of skilled labourers and gunny bags for grading and packing. There was lack of vehicles at the desired time. Lack of metalled and all weather roads was also reported.

Regarding market intelligence's wrong and misleading information were made available to potato growers, various mal-practices were prevalent part payment and forced low payment were also reported lack of technical know how was also reported. Suggestions of above mentioned problems are as follows:

1. Processing units must be developed on or nearby the mango and potato farms to solve the problems of unemployment and disguised employment among growers.
2. Encouragement's by the government should be facilitated to farmers for covering more and more area under mango and potato as there are most profitable crops.
3. Cropping pattern should be changed in such a way so that mango and potato growers can allot more area under these crops.
4. Mango growers must be free from the clutches of contractors who grabe their margin and profit.
5. Marketing infrastructures for mango must be ameliorated.
6. Number of cold storage's should be increased to save the wastage of potato.
7. Harvest prices of both mango as well as potato should be increased to encourage growers and there must be correct pricing policies.
8. Problem of erratic electric and diesel supply must be finished.
9. Along with transportation grading and packing materials must be supplied at cheaper rates.
10. Technical know-how and correct market intelligence should be provided to growers.

WEST BENGAL

The nature and extent of problems/constraints in producing the horticultural crops as perceived by the sample farms have been thoroughly assessed in this section. There are seven field level problems/constraints according to the perception of the farmers which play as a powerful barrier in production of horticultural crops in general and the orchid in particular. However, the degree of intensity of these constraints vary from farm to farm.

Orchid in particular and the horticultural crops in general being an export potential crops of the state, the prosperity of the growers as well as the overall economy of the state is closely interlinked. The result of the present study clearly demonstrates that there is a promising trend towards cultivation of these crops throughout the state because these crops are not only profitable in terms of higher output-input ratio but these crops also generate a higher wage employment for agricultural sector.

However, for adequate growth of this sector vis-a-vis expanding of export of these products attention may be made in developing appropriate organisation. In this juncture a change in the land ceiling act may be amended for allowing the corporate companies to invest more capital on development of large sized captive gardens. There is also an urgent need to adopt effective steps to fulfil the objectives of transfer of technology (TOT) in a proper manner. In this connection, for effective agricultural extension, a strong need is felt for a closer co-ordination and monitoring between the Agricultural Directorate and Horticultural directorate of the state. Agricultural University can play more effective role in extension and development activities, as well as in transferring the latest technologies at the door step of the growers.

11

WTO ISSUES

Trade and markets for food and farm goods matter to us all — whether we are consumers of these products, or also produce, transform and trade them.

To address government policies that distort markets and restrict trade, WTO members concluded the Agreement on Agriculture, which came into force in 1995. This initiated reductions in subsidies and trade barriers to make markets fairer and more competitive. It also provided for members to continue negotiations for further reform, taking into account concerns such as food security and the environment. These ongoing talks led in 2015 to a historic decision to abolish agricultural export subsidies and new rules for other forms of farm export support.

The WTO's Agriculture Committee oversees implementation of the Agreement and provides a forum for members to address related concerns.

The World Trade Organization (WTO) is an intergovernmental organization that regulates and facilitates international trade. Governments use the organization to establish, revise, and enforce the rules that govern international trade. It officially commenced operations on 1 January 1995, pursuant to the 1994 Marrakesh Agreement, thus replacing the General Agreement on Tariffs and Trade (GATT) that had been established in 1948. The WTO is the world's largest international economic organization, with 164 member states representing over 98% of global trade and global GDP.

The WTO facilitates trade in goods, services and intellectual property among participating countries by providing a framework for negotiating trade agreements, which usually aim to reduce or eliminate tariffs, quotas, and other restrictions; these agreements are signed by representatives of member governments :fol.9–10 and ratified by their legislatures. The WTO also administers independent dispute resolution for enforcing participants' adherence to trade agreements and resolving trade-related disputes. The organization prohibits discrimination between trading partners, but provides exceptions for environmental protection, national security, and other important goals.

The WTO is headquartered in Geneva, Switzerland. Its top decision-making body is the Ministerial Conference, which is composed of all member states and usually convenes biennially; consensus is emphasized in all decisions. Day-to-day functions are handled by the General Council, made up of representatives from all members. A Secretariat of over 600 personnel, led by the Director-General and four deputies, provides administrative, professional, and technical services. The WTO's annual budget is roughly 220 million USD, which is contributed by members based on their proportion of international trade.

Studies show the WTO has boosted trade and reduced trade barriers. It has also influenced trade agreement generally; a 2017 analysis found that the vast majority of preferential trade agreements (PTAs) up to that point explicitly reference the WTO, with substantial portions of text copied from WTO agreements. Goal 10 of the United Nations Sustainable Development Goals also referenced WTO agreements as instruments of reducing inequality. However, critics contend that the benefits of WTO-facilitated free trade are not shared equally, citing the outcomes of negotiations and data showing a continually widening gap between rich and poor nations.

World Trade Organisation (WTO) provides a legal and institutional framework for a multilateral trading system. It is basically committed to establishing an open and liberal global economy, free from any barrier or restriction on trade, and to encourage participation of both developed and developing countries in the multi-lateral trading system. Under this, there is also a desire for making efforts to ensure that the developing countries and least developed countries secure a share in international trade. The basic principles, which are of importance and relevance to horticultural exports under WTO, are discussed in this chapter:

PRINCIPLES OF THE TRADING SYSTEM

The WTO establishes a framework for trade policies; it does not define or specify outcomes. That is, it is concerned with setting the rules of "trade policy." Five principles are of particular importance in understanding both the pre-1994 GATT and the WTO:

1. Non-discrimination. It has two major components: the most favored nation (MFN) rule and the national treatment policy. Both are embedded

in the main WTO rules on goods, services, and intellectual property, but their precise scope and nature differ across these areas. The MFN rule requires that a WTO member must apply the same conditions on all trade with other WTO members, i.e., a WTO member has to grant the most favorable conditions under which it allows trade in a certain product type to all other WTO members. "Grant someone a special favor and you have to do the same for all other WTO members." National treatment means that imported goods should be treated no less favorably than domestically produced goods (at least after the foreign goods have entered the market) and was introduced to tackle non-tariff barriers to trade (e.g. technical standards, security standards et al. discriminating against imported goods).

2. Reciprocity. It reflects both a desire to limit the scope of free-riding that may arise because of the MFN rule and a desire to obtain better access to foreign markets. A related point is that for a nation to negotiate, it is necessary that the gain from doing so be greater than the gain available from unilateral liberalization; reciprocal concessions intend to ensure that such gains will materialize.

3. Binding and enforceable commitments. The tariff commitments made by WTO members in multilateral trade negotiation and on accession are enumerated in a legal instrument known as a schedule (list) of concessions. These schedules establish "ceiling bindings": a country can change its bindings, but only after negotiating with its trading partners, which could mean compensating them for loss of trade. If satisfaction is not obtained, the complaining country may invoke the WTO dispute settlement procedures.

4. Transparency. The WTO members are required to publish their trade regulations, to maintain institutions allowing for the review of administrative decisions affecting trade, to respond to requests for information by other members, and to notify changes in trade policies to the WTO. These internal transparency requirements are supplemented and facilitated by periodic country-specific reports (trade policy reviews) through the Trade Policy Review Mechanism (TPRM). The WTO system tries also to improve predictability and stability, discouraging the use of quotas and other measures used to set limits on quantities of imports.

5. Safety values. In specific circumstances, governments are able to restrict trade. The WTO's agreements permit members to take measures to

protect not only the environment but also public health, animal health and plant health.

There are three types of provision in this direction:

1. articles allowing for the use of trade measures to attain non-economic objectives;
2. articles aimed at ensuring "fair competition"; members must not use environmental protection measures as a means of disguising protectionist policies.
3. provisions permitting intervention in trade for economic reasons.

Exceptions to the MFN principle also allow for preferential treatment of developing countries, regional free trade areas and customs unions. :fol.93

BASIC PRINCIPLES

Non Discrimination

A country should not discriminate its trading partner i.e. every member country will be granted the same status that is granted to Most Favoured Nation. It means that products made in member's own country are not given any favourable treatment in comparison to the products which originated from other member countries and are imported into their territory. It includes that the procedures and standards which are applicable to the domestic industry will automatically be implied to the imported goods i.e., once a foreign good satisfies the conditions including payment of customs duty and/or other charges, it will be treated no less favourably in terms of taxes and measures with equivalent effect than domestic goods.

Reciprocity

The agreements administered by WTO are based on a balance of rights and obligations achieved through reciprocal exchange of market access commitments. The open trading system is to be based upon the rules for bringing competition.

Market Access

An important goal of the WTO is to facilitate the availability of the goods across the border. This agreement also helps in increasing the transparency under the trade regimes. Through this Agreement, the trade accross the borders is only governed by imposition of tariffs while rest of the quantitative restrictions, such as quota are to be waived off in a phased manner. In order to exploit the marketing accessibility, competitiveness of a product with respect to its price and quality is a fundamental criteria.

Fair Competition

The main objective of this principle is to discourage the unfair trade practices such as export subsidies and dumping of products at below cost to gain market share. It is being carried out through regulation of subsidies on export and imposition of anti-dumping duties.

AGREEMENT ON AGRICULTURE

Agreement on Agriculture has 4 main components, namely: -

- Concession and commitments on market access
- Domestic support and export subsidies
- The Agreement on sanitary & phytosanitory measures, and
- Ministerial declaration concerning least developed and under developed countries.

The Agreement on Agriculture seeks to liberalise the world trade in agriculture so as to make it free from Government measures that inhibit the trade movement and thereby inefficiency of the sector. Under this agreement, there are different ways of subsidising farmers through governmental interventions. As per this agreement, the member countries are required to reduce the direct subsidies i.e. the subsidies which are based on products or are being given to the individual farmers either for the promotion of export or selling their products at a price below the cost of supply.

However, the other subsidies which are also called green subsidies are presently being allowed under this agreement. The allowed subsidies are basically broad based in terms of research and pest and disease control, training services, extension services, marketing and promotion services, infrastructural services in order to exploit the market accessibility.

After going through the agreement, it clearly comes out that in order to exploit the facilities being given in the above agreement, there is need to overcome the infrastructure inadequacy and other impediments to the large scale investments. Apart from this, it is necessary that post-harvest technologies for food processing and preservation should be promoted so as to bring about added value to the agriculture economy.

Finally, the gain or loss to a country as a result of liberalisation will largely depend on achieving the market accessibility through providing the green box subsidies.

Indian export basket consisting of agro-based products and manufacturing items suffers from considerable supply related problems. So far as the agro based products are concerned, the question of surplus is most critical. In case of manufacturing products, there are serious problems of competition relating to high cost, inefficiency in production, outdated technology, problems related to delivery schedule and product specification.

Categorization of Rubber

While coffee and tea have been classified under 'agriculture' and come under the purview of the 'Agreement on Agriculture', natural rubber has been taken as an industrial raw material and does not come under this Agreement. This implies that natural rubber will not enjoy tariff and non-tariff protection generally available to agricultural commodities. Rubber is a tree product and is practically grown in the rain-fed plains, in elevations up to about 1500 feet.

There are nearly one million growers of rubber, cultivating the crop in about 0.55 million hectares in the country. Most of them are small farmers with an average holding size of less than 0.5 ha. Their livelihood is solely dependent on the returns from the crop. They tap the rubber trees, collect the latex, dry it as sheets and sell in the market. There is no manufacturing process involved as far as the growers are concerned and the entire operation is purely agricultural. In fact, the entire income from natural rubber in sheet form is treated as agricultural income under the Indian constitution and is taxed accordingly under the Agricultural Income Tax Acts of the respective state governments. In these circumstances, there is no justification to exclude natural rubber from the purview of agriculture. It is pertinent to note here that natural rubber has been classified as agricultural raw material under the commodity trade classification of UNCTAD as well as under the Integrated Programme for Commodities (IPC) of the Common Fund for International Development Assistance.

There are several disadvantages arising from rubber not being included in the category of 'agriculture' under WTO. The first disadvantage is the bound rate i.e., the maximum rate of import duty, allowed to be levied. It is noted that the bound rates for agricultural commodities are relatively higher. While the bound rate for tea is 150%, and coffee and pepper 100%, that for natural rubber is as low as 25%. Another aspect is that agricultural commodities come under the basket of 'green box', arising from which certain concessions including developmental subsidies could be extended and we may have certain bargaining powers on the ground that items in this box are eco-friendly in contrast to pollution-prone industrial products.

In view of the above:

- At the time of the next round of WTO negotiations, India may take up that, in all fairness, natural rubber should be classified as an 'agricultural commodity', and should be brought under the purview of the Agreement on Agriculture, which is fully justified.
- Meanwhile, the bound rate on natural rubber should be increased to a reasonably higher level to be on par with other agricultural products. It should be examined if this could be done through the provisional safeguard measures, in the wake of the QR removal.
- Rubber is a tree crop which is environmentally friendly and therefore, be put under the 'green box'.

AGREEMENT ON SANITARY AND PHYTO-SANITARY (SPS) MEASURES

The objective of this Agreement is to improve health or life of human, animal and plant and overall phyto-sanitary situations in all the member countries. This agreement states, that the measures should not be applied in a manner, which would result in a means of arbitrary or unjustifiable discrimination between the members. In other words, it means that the sanitary and phyto- sanitary measures applied by each member country should be based on scientific justification and to the extent possible, it should take into consideration, the international standards and guidelines already available. The major articles in the Agreement having an effect on the horticulture produce are as follows

Article 3 on Harmonisation

This Article states to harmonise the sanitary and phyto-sanitary measures, which basically include levels of food additives, contaminants, pesticide residues, microbiological criteria and other related toxins with the international standards and guidelines. It means that the standards already available in India for fresh horticulture produce should be immediately harmonised with the international guidelines and if higher level of measures is required, the scientific justification for the same may be documented at the earliest. This process of harmonisation is crucial with respect to both exports and imports. In case our domestic standards are inferior to the international guidelines, we will not be in a position to impose stringent standards on the imported products, thereby, all the inferior quality products will be dumped into our country. This issue of harmonisation is also required when we enter into an equivalence agreement for exporting our goods into other countries. Before entering into equivalence, we are required to submit the detailed data on the SPS measures applied within our country and the mechanism for monitoring the implementation of the same.

Article 4 on Equivalence

Article 4 of the Agreement directs the WTO members to accept SPS measures of the other countries as equivalent even if these measures differ from their own or from those used by other countries trading in similar products, provided, the supplier country can demonstrate to the importing country that its measures achieve the importing country's appropriate level of SPS protection. Under this Article, WTO members can also enter into bilateral or multi-lateral agreements on mutual recognition of equivalence of specified SPS measures. The current trend is that many developed countries are entering into mutual equivalence agreements as it facilitates trade between the respective countries. This has been one of the reason for a more favourable treatment of products imported from developed countries as against products originating from developing countries resulting in the latter's further marginalisation.

Article 5 on SPS Measures to be Based on Scientific Evidence

As per Article 5 of the Agreement, countries are required to ensure that the measures of SPS protection are based on risk assessment on the basis of available scientific evidence, production processes and environmental conditions. Where scientific evidence is lacking, countries can resort to temporary SPS measures. In such cases, countries are required to obtain scientific evidence for a final position and within a reasonable time frame. Moreover, such SPS measures cannot be more trade-restrictive than is necessary to achieve the appropriate level of protection taking into account the technical and economic feasibility.

Article 7 on Transparency

According to Article 7 (transparency), all WTO members have an obligation to notify to the WTO committee, any changes in their SPS measures. They are also required to notify measures, which have a "significant effect on trade of other Members" and that depart from an international standard guideline or recommendation.

Article 9 on Technical Assistance

As per Article 9, the WTO members are expected to facilitate technical assistance to other members especially the developing countries either bilaterally or through the relevant international organisations. This assistance may be in the areas of processing technologies, research and development, infrastructure, etc. and may be in the form of information, credits, grants, training and equipment to enable the countries to comply with the SPS measures necessary to achieve the level of SPS protection adopted by the importing country.

Article 10 on Special & Differential Treatment

As per Article 10 of the SPS Agreement, Special and Differential Treatment (SDT) is required to be provided to the developing countries, whereby, longer time frames for compliance with SPS measures are supposed to be provided by the developed countries. Moreover, the SPS Committee can grant developing countries specified time limit exceptions to comply with the application of the SPS Agreement.

PRODUCTS FACING SPS RELATED PROBLEMS

Mango and Grape

Japan, Australia and China have banned imports of mangoes and grapes from India on account of presence of certain fruit flies. China has imposed a ban on grapes for a species of fruit fly that does not even exist in India. Australia desires to have complete details about pest management practices in India and a ban can only be lifted after signing of a MOU on mutual recognition of pest management practices. US rules governing import of fresh fruits and vegetables are very stringent. USDA gives clearance only after detailed tests involving inspection of the production areas.

Japan, on the other hand, desires Vapor Heat Treatment (VHT) of fruits for dis-infestation before these can be allowed to enter into their country. Lot of research work has been carried out in India to meet the Japanese requirements. It has taken more than five years to develop the protocols and inspite of this lifting of the ban is not anticipated in the near future. The lengthy procedure involved is that even after evaluation of our report (which itself is pending for one year), the Japanese quarantine experts have yet to visit India for their own evaluation on the VHT machine. Thereafter, a public debate will have to be carried out in Japan before the decision on lifting of the ban can be considered. In the process, we are likely to lose a total of about 10 years before India can think of actually exporting these fruits to Japan. The Japanese regulation is too strict a SPS measure for any country to comply with.

It is, informally learnt that these restrictions are actually a non tariff barrier to trade because of the commercial interest of Japanese companies in Philippines and Thailand, where they have their own mango orchards for meeting the Japanese demands.

Coffee

The European Union (EU) introduced a regulation prescribing unreasonably low levels of Octratoxin-A (OTA) in coffee. The implementation of the EU regulation got delayed after 38 supplier countries presented a joint memorandum to the EU stating that the proposed EU regulation was not strictly related to the extent of risk involved and it also had negative trade effects. The EU has decided to review the situation by the end of 2002.

PESTICIDE RESIDUE PROBLEMS

Grape

The European Union (EU) has introduced regulations on pesticide residue levels in fruits and vegetables imported from different countries. A number of chemicals have been listed and more are being added on a regular basis. Indian exporters are required to comply with the maximum pesticide residue levels for exporting their products to the EU. This problem has specially been faced in exporting grapes, where a number of procedures had to be introduced by India to ensure that grapes were exported with pesticide residue levels within the prescribed limits.

Gherkins

India is currently the largest exporter of provisionally preserved gherkins to the EU. Our exports, however, have shown a declining trend during the last two years. One of the reasons is different pesticide residue levels fixed by different member states of the EU. Moreover, the chemicals for which residue levels have been prescribed also differ from country to country within the EU with the result that there are no harmonized regulations in this regard and no intra-EU trade in gherkins. This is restricting free flow of trade in this commodity.

Tea and Spices

There are also different regulations of the EU member countries with regard to import of spices and tea, whereby maximum pesticide residue levels prescribed by them have to be complied with India is particularly facing problems in export of chilli to Spain which has different pesticide residue levels as compared to those of another EU country.

Honey

The EU has prescribed maximum levels of pesticide residues in honey and these are required to be tested before exports to the EU. While this appears to be an appropriate level of protection, the procedure of implementation is cumbersome. India is required to provide to the EU with its residue plans for the previous year and for the following year with respect to honey produced in all regions of the country. This is practically difficult keeping the Indian conditions in view where honey is required to be collected from a large number of producers different sources before being packed. One may not be surprised if India sees a ban on exports of honey to the EU sooner than later.

FACTORS GOVERNING COMPETITIVENESS

Lack of Range of Varieties

While a wide variety of germplasm is available in most of the horticultural commodities, varieties most suited for exports have not been adequately identified in several crops. Similarly, many well-known varieties having demand in the international market have not been introduced and tried. Work on developing export -oriented varieties and those suited for value addition needs to be given due priority

Pre- Harvest Practices to Control Post Harvest Losses

Several post harvest diseases and pests can be avoided with appropriate pre-harvest treatment. Such recommendations need to be popularized among growers to reduce losses and improve quality. Apart from this, pest infestation is also a problem, which is affecting the price realization of Indian produce in the market. Both these problems can only be dealt with by following the correct pest management and pre- harvest practices.

Loss of Produce at the Primary Level

There is significant loss of produce at the primary production level as a result of inappropriate harvesting methods. There is therefore need to develop and implement proper harvesting methods. An important example is mango, which is required to be plucked at a stalk length of more than 5 cms in order to prevent the fungus infection. Once the stalk is removed either during or soon after, the sap flows on the fruit surface spoiling its appearance and often resulting in fungus infection. This problem could be minimized to a great extent by using the mango harvester, which has been developed in the country in a number of institutions e.g. IIHR Bangalore, CISH Lucknow and KKV Dapoli. Efforts would be needed to popularize the same among the farmers.

Lack of Adherence to Maturity Indices

The maturity indices in most of the horticultural crops are either not available or not followed by the farmers resulting in harvesting of these crops at varied maturity levels. Thereby, a large quantity of produce has to be rejected while sorting and grading due to under or over maturity/ripening e.g. fruits and vegetables. Further, once such a lot is exposed to controlled atmosphere condition or modified atmosphere packaging, the reaction of the commodity is varied, making it difficult to standardize the post-harvest treatments. In crops like coconut and oil palm lack of implementation of uniform maturity standards can result in avoidable loss in quantity and quality. Hence, there is need to create awareness about the maturity indices of different horticultural commodities among the farmers both for catering the needs of export market and for marketing of produce in distant locations within the country.

Lack of Facilities for Physical and Chemical Treatment After Harvesting

Several post harvest treatments are recommended for minimising post harvest losses in different horticulture commodities. However, adequate facilities for large-scale treatment are not available at the farm level. As a result, gap between harvesting and post-harvest treatment such as hot water dip is very long resulting in avoidable losses to quality.

Lack of Post Harvest Infrastructure and Logistics

In order to maintain the quality of horticultural commodities there is need to develop proper infrastructure in terms of pre-cooling units, grading and packing houses in the producing areas. For instance, in Controlled Atmosphere (CA) storage, the time gap permitted between the harvesting and the hot water dip is 8 hours. However, due to inadequate packing house facility and lack of refrigerated transport, the above condition is difficult to implement in practice. Such a situation is not conducive to promoting exports. In view 3of this, it is the need of the hour to establish pre-cooling units or packhouses as well as cold storages for fruits, vegetables and flowers in the growing regions. Under WTO, the subsidies for creation of such an infrastructure are allowed. Therefore, the

Government should take immediate steps for creation of such facilities for the common use of all the farmers in different production regions of the country.

NEW OPPORTUNITIES

Organic Farming

There is a need to promote organic farming in production of different horticultural commodities with the objective of promoting exports and improve returns. There is, therefore, an urgent need to develop technology for organic farming and create awareness on the benefits it provides. This calls for standardization of technologies for producing vermi-compost, biofertilizer, mycorrhizae and bio pesticides for different crops. There is also need to develop guidelines for this purpose and designate agency / agencies for accreditation of such produce meeting international requirements.

Import Substitution

There is vast scope of reducing imports of horticultural commodities like dates, nuts like almond, pecannut, pistachionut, macedemia nut, palm oil, raisins, cocoa and rubber. Area under such commodities needs to be increased. There is also need to develop commercial plantations of fruits like mangosteen, durian, longan,rambutan, etc which are commercially grown in most south Asian countries namely Thailand, Malaysia and Indonesia.

Import Intimation System

The need for having adequate, reliable and timely data in respect of import of commodities has become extremely necessary now in the context of opening up of the economy. There should be on line records about quantity, value, quality etc., of the commodities imported in the country so that both the government and interested entrepreneurs could make use of this information for various purposes. There is more urgent need for commodities like spices, cashew where international trade is sizeable.

Products having an Edge in International Market

India being largely a tropical country, it is necessary to concentrate efforts on tropical horticultural crops such as fruits like mango, litchi, sapota, passion fruit, aonla, bael and guava and plantation crops like coconut, cashew, cardamon, black pepper, tea and coffee. Since most of the developed and importing countries are in the temperate region the chances of increasing our market share with such crops in this region are more. Apart from this, India is also very rich in medicinal plants. In order to promote export, R&D on these crops needs to be intensified and cultivation of crops with export potential needs to be promoted. In case of coffee, there is a need to exploit the potential of Indian Arabica Gourmet Sectort and of robusta in expresso segment.

PRODUCT SPECIFIC MEASURES

Floriculture

Cut flower industry in India is at a nascent stage. In recent years floriculture units have come up near most metropolitan cities like Bangalore, Pune, Delhi, Hyderabad and Chennai. India enjoys the advantage of favourable agro- climate, cheap labour, extensive arable land and skilled manpower. Because of these reasons, India has a very high potential for export of cut flowers. The Government of India has given Thrust Area Status for floriculture industry as a potential foreign exchange earner.

Floriculture exports from India have shown a steady growth and have increased from Rs 146 million in 1990-91 to Rs 1733.7 million in 1998-99. Among Floriculture products, the growth of fresh cut flower sector has been very significant. From Rs 10.2 million in 1991-92 to Rs 251.1 million in 1998-99, this cumulative growth of about 90% has come largely due to the proactive role played by APEDA in promoting this industry. However, India's share in the global cut flower trade remains less than one per cent. This shows the potential and the vast opportunities that exist for export of cut flowers. With new markets being identified and exporters having negotiated the learning curve, the prospects are improving.

Rose, orchids, anthuriums, carnations, chrysanthemums and gerbera are the most common cut flowers having a major share in the international trade.The key flower exporting countries are Holland, (59% share in the world export market), Columbia (10%), Italy (6%) and Israel (4%). The major destination for export of cut flowers from India is Europe (45%), Japan (30%), Australia (10%) and other countries (15%). There is thus a good potential for export of flowers from India to Europe, Japan, Australia, Singapore and North America.

HORTICULTURAL QUALITY STANDARDS

Indian Standards In Agricultural Products

Indian standards in the agricultural sector have been framed by various organisations of the Government most of which are under Ministry of Agriculture. These organisations are responsible not only for production and product standards, but also for their inspection and quality control. Some of the major organisations involved are given below in Table 4.1

Agreement on Technical Barriers To Trade

This Agreement deals with the product standards related to physical parameters such as size, colour, appearance, maturity, packaging and labelling requirements. Presently, for most of the horticulture produce the standards stating the above requirements are not mandatory for domestic produce, in lieu of which, the imported products are enjoying the liberty of following their own standards as applicable in their country. As a result several of the packaged imported products are having labels which are in foreign languages. However, as per our national legislation on packaged item it should be made incumbent on all packaged products to have labels either in English or Hindi.

Apart from the product standards, there is also need to develop procedure for implementation and monitoring of these standards from import point of view. As per the above Agreement, all the standards should be clearly laid down so that they are published in a transparent manner and other member countries may utilise these in order to promote their exports. In order to meet these conditions, there is an urgent need to specify procedure in this regard.

Harmonisation of Standards

There is also multiplicity of standards in several horticultural products. There is, therefore, an urgent need to not only rationalise standards fixed by various organisations but also to harmonise these with ISO standards for different commodities. Standards for growing and package requirements of international markets are also not available in a large number of commodities. The standards developed by Directorate of Marketing and Inspection are old and outdated. There is, therefore, an urgent need to fix standards to promote domestic as well as international trade in agricultural / horticultural commodities.

INTELLECTUAL PROPERTY RIGHTS (IPR)

In the present scenario of globalisation, it has become important to understand the implications of the IPR. A patent is a legal monopoly granted to the owner of any new invention, which is capable of being used for limited period of time. It is a privilege granted by the Government to an inventor and other persons deriving their rights from inventions. The patents also stimulate the technology process through diversification of products and upgradation of the technologies. The owner of a patent on a product / variety has full right to prevent others from making, using or importing this product without his consent. Presently, in India there is no legislation, which covers the protection of plant breeders right. As a result, many of the foreign nurseries do not encourage the supply of horticultural crop varieties to India since their rights are not fully protected. Thus, India does not have access to many useful materials due to non-protection of plant breeder's right.

QUALITY ISSUES

In the light of recent changes in the international trade, to exploit the potential for export of horticulture produce to its full capacity, sincere efforts are required to be made to develop a full fledged Quality Assurance Machinery. This proposal would affect business of all sizes engaged in production and trade of products and services throughout the supply chain. In addition to compliance with national legislation, this machinery will ensure that the following issues are considered.

- To ensure that all stages of production and distribution for which they are responsible are carried out in such a manner as to comply with food safety and consumer protection requirement;
- To inform the competent authority if it considers or suspects that a product placed on the market does not comply with the relevant food safety requirement and of action taken to prevent risk to the final consumer;
- That no product is exported which does not comply with the general requirement of the food laws;
- To have a system in place to identify the supplier of the product and the buyers or traders to whom they have supplied such products.
- To adequately label and identify produce to allow its traceability and withdraw the product if it considers or suspects that it is not complying with food safety requirements.

The above machinery would also be useful for ensuring that the products imported into India comply with conditions outlined above. In the present situation, there are chances that the developed countries could dump their products at throw away price to capture the Indian market along with the pests and diseases. Similarly it will ensure that genetically modified organisms (GMO) and microorganisms presently not existing in Indian soil do not enter into the food chain of the country.

There is a need to have a national approach for ensuring the quality and safety of food items including domestic, imported and exported products. This necessitates the overhauling of the current monitoring and control mechanism to ensure that the products entering into Indian market comply with the national food safety laws. In order to give a systematic approach to this important issue, a two-way action plan is suggested.

Quality Building Machinery

Includes activities targeted to upgrade the status of food processing industries including handling of fresh produce to achieve good quality and safe food products, through creating awareness and training among food industries. This has to be ensured by the industry.

Quality Assurance Machinery

To ensure development and implementation of produce and process standards through monitoring and certification thereby ensuring the quality and safety of the final product.. This is primarily the role of the Government in the current scenario.

LAYING DOWN STANDARDS

The task of laying down the standards, their notification and ensuring their adherence may be divided within different Government Departments product wise such as:

Voluntary Standards

A National Committee for Product and Process Standard should be formed under the umbrella of Ministry of Agriculture. Its role would be to form different subject wise-sub committees under various Departments of the Ministry of Agriculture. The other members of the Sub-Committee may include various promotional boards and state departments. These sub-committees may be constituted product-wise such as: -

Sub-Committee on Fresh Fruits and Vegetables

This Committee may be formed under the chairmanship of Directorate of Marketing and Inspection for developing Standards for fresh fruits and vegetables. These standards should include the following requirements: -

- Name of the standard
- Scope
- Description
- Essential composition and quality factors

- Food additives and contaminants
- Hygiene
- Weights and measures
- Labeling
- Method of analysis and sampling

In order to achieve the goal of safety of the product, the final product testing is not a reliable criteria. In view of this, it is required that the practices, which are implemented at the production and harvesting level should also be improved. In view of this, process standards should be developed which would identify the critical points, which are required to be monitored at the production and the harvesting level. It would also provide reference in terms of guidelines for production of fresh fruits and vegetables during primary production and packaging.

Sub-Committee on Processed Products

Processed Products would include Processed Fruits and Vegetables, Nuts, alcoholic and non alcoholic beverages, coconut, spices, tea, coffee, honey, medicinal & herbal preparations, etc. Product wise Sub-Committees should be formulated to develop product and process standard under the umbrella of Ministry of Agriculture i.e. Department of Food Processing Industries. These standards should also be harmonized in terms of format prescribed under international standards such as Codex Standards. The different Product Orders such as FPO specification which lays down requirements related to location of the factory, personnel hygiene, equipment requirements, and its maintenance etc. are also required to be modified in order to incorporate the hygienic practices as prescribed under Codex i.e. General Principles for Food Hygiene.

Mandatory Standards

The standards related to Sanitary and Phyto Sanitary (SPS) aspects listed below are refereed to as Mandatory standards:

- Minimum residue level for pesticides (MRL's)
- Heavy metals
- Mycotoxins
- Microbiological requirements
- Any other requirements related to food safety of processed horticultural products.

All the requirements may be notified under PFA (Prevention of Food Adulteration Act) because the major objective of PFA formulation is to protect the health of the consumer. Thereby, the rule formulated under PFA would be mandatory for all the commodities for domestic consumption import and export.

Monitoring and Inspection Mechanisms to Ensure Implementation of Product and Process Standards

Once the product and process standards for various commodities are laid down and notified under the various departments, the next aspect which requires attention is the monitoring and inspection of the above standards. Primarily the monitoring and inspection is the role of the government. However, this task may be carried out through accreditation of inspection agencies which would be accredited the task of carrying out the inspection on behalf of the government before granting approval for processing of the respective food commodities. In order to achieve this objective of accrediting the inspection bodies which may include both private and government bodies, the following action is required to be taken

- Laying down the guidelines for assessment and accreditation of inspection and certifying bodies which would also cover the legislative framework and other infrastructure requirements including control programmes.
- Laying down the guidelines to be followed by inspecting bodies for inspection and monitoring of food processing industries for certification.

The responsibility of ensuring the implementation of above product and process standards and also laying down the guidelines for assessment, accreditation and inspection would lie with the Departments which are chairing product wise Sub-Committees. These Departments may also accredit private agencies which would further identify the certification bodies to carry out the inspection of food industries for compliance to the product and process standards on behalf of each Department.

The overall evaluation of the situation related to the implementation and assurance of product and process standards may be with National Committee of Product and Process Standards. It would include the yearly review of the status of national standards with respect to the changes taking place in international scenario related to the aspects of quality and safety in food trade.

Networking

In the light of the above developments, a networking of the laboratories also requires equal importance as these laboratories would be utilized for analysis to ascertain whether the products comply with the standards laid down by the government. It is, therefore, necessary that there is a wide network of laboratories distributed through out the country to carry out this task. In order to monitor this task, a sub-committee under the National Committee may be formed which would recognise laboratories as per the standards, laid out. This would primarily include the following:

i) Standards related to the various technical capabilities required with respect to the target product to be tested.

ii) The system requirements specifying the procedures for processing of samples and other documentation requirements needed to comply with for recognition of the laboratory.

iii) The role of this Committee will be to develop a system of recognition and continuous surveillance of the laboratories. In order to ensure the implementation of above standards, the whole task may be done through a nodal agency which is presently National Accreditation Board for Laboratories under the Department of Science and Technology. This agency has also notified standards for laboratories including detailed procedure for accreditation of these agencies at the government and non-government level for recognising the laboratories meeting the above-notified standards. A regular feed-back mechanism is also needed to be in place, in order to continuously improve the procedure where necessary.

Nodal Agencies

With the approach of introducing system standards within the food industries such as Hazard Analysis Critical Point (HACP) and ensuring their implementation through the certification process, it is required that a nodal agency may be identified, with following role.

- Laying down the guidelines for implementation of food safety management system in food industry including standards for HACP based on the Codex requirements.
- Laying down the guidelines for assessment and accreditation of certification bodies for the quality assurance system.

BIBLIOGRAPHY

- Andrew H. Cobb; John P. H. Reade (2011). 7.1. Herbicides and Plant Physiology. John Wiley & Sons. ISBN 9781444322491.
- Anonymous, 1994, Horticultural Growth: Abundant Products, Economic Times, June, 25 P 12, F.A.O.
- Dave, V.J., 1998, Economics of Export Oriented Horticultural Crops Chiku (Sapota) in Gujarat (A case study in Valsad district), Agro-Economic Research Centre, Sardar patel University Vallabh Vidyanagar, Gujarat, 1998, (Memio).
- EPA. February 2011 Pesticides Industry. Sales and Usage 2006 and 2007: Market Estimates Archived 2015-03-18 at the Wayback Machine.. Summary in press release here Main page for EPA reports on pesticide use is here.
- FAO, 1994, year Book on Trade & Commerce, Vol. 148, 1994, F.A.O.,Rome.
- Jagadeswara, M. Rao, 1998, Economics of Export Oriented Horticultural Crops in Andhra Pradesh, Agro-Economic Research Centre, Andhra University, Visakha Patnam March, 1998 (Memio).
- Jha, U.M., 1997, Economics of Export Oriented Horticultural Crop (Litchi) in Bihar, Agro-Economic Research Centre for Bihar, T.M. Bhagalpur University, Bhagalpur, Sept., 1997(Memio).
- Kaul, G.L., 1997, Horticulture in India - Production Marketing and Processing, Indian Journal of Agricultural Economics, Vol. 52, No.3 July-Sept., 1997 pp 56.
- Krämer, Wolfgang, ed. (2012). Modern crop protection compounds (2nd, rev. and enl. ed.). Weinheim: Wiley-VCH-Verl. pp. 197-276. ISBN 978-3-527-32965-6.

- Lock, E. A.; Ellis, M. K.; Gaskin, P; Robinson, M; Auton, T. R.; Provan, W. M.; Smith, L. L.; Prisbylla, M. P.; Mutter, L. C.; Lee, D. L. (1998). From toxicological problem to therapeutic use: The discovery of the mode of action of 2-(2-nitro-4-trifluoromethylbenzoyl)-1,3-cyclohexanedione (NTBC), its toxicology and development as a drug. Journal of Inherited Metabolic Disease. 21 (5): 498-506. doi:10.1023/A:1005458703363. PMID 9728330.
- Malik, R.P.S., 1998, Economics of Export Oriented Horticultural Crops -- A case study of Floriculture in Haryana, Agricultural Economics Research Centre University of Delhi, Delhi, August, 1998, (Memio).
- Mallory-Smith, C. (1999). Impact of labeling herbicides by site of action: A University view. Weed Technology. 13: 662
- Moran GR (Jan 2005). 4-Hydroxyphenylpyruvate dioxygenase (PDF). Arch Biochem Biophys. 433 (1): 117-28. doi:10.1016/j.abb.2004.08.015. PMID 15581571. Archived from the original (PDF) on 2014-03-03.
- Naganathan, M.,1998, Economics of Export Oriented Horticultural Crops in Kerala, Agro-Economic Research Centre, University of Madras Chennai, 1998 (Memio).
- Protecting Crop Yields And Quality Worldwide.
- Quastel, J. H. (1950). 2,4-Dichlorophenoxyacetic Acid (2,4-D) as a Selective Herbicide. Agricultural Control Chemicals. Advances in Chemistry. 1. pp. 244-249. doi:10.1021/ba-1950-0001.ch045. ISBN 978-0-8412-2442-1.
- Rahim, Kazi M.B. and Debas his Sarkar, 1998, Economics of Export Oriented Horticultural Crops of West Bengal, Agro-Economic Research Centre, Visva-Bharati, Santiniketan, 1998(Memio).
- Rahim, Kazi, M.B. and Debashis Sarkar, 1998, Economics of Export oriented Horticultural Crops in Sikkam, Agro-Economic Research Centre, Visva-Bharati, Santiniketan, 1998 (Memio).
- Retzinger Jr, E. J.; Mallory-Smith, C. (1997). Classification of herbicides by site of action for weed resistance management strategies. Weed Technology. 11: 384-393.
- Robbins, Paul (2007-08-27). Encyclopedia of environment and society. Robbins, Paul, 1967-, Sage Publications. Thousand Oaks. p. 862. ISBN 9781452265582. OCLC 228071686.
- Robert L Zimdahl (2007). A History of Weed Science in the United States. Elsevier. ISBN 9780123815026.

- Saikia, T.N., D.Borah and Bharti: Gogoi, 1997, Economics of Export Oriented Horticultural Crop (Pine apple) in Tripura, Agro-Economic Research Centre For North East India, assam Agricultural University Jorhat, 1997(Memio)
- Schmidt, R. R. (1997). HRAC classification of herbicides according to mode of action. 1997 Brighton crop protection conference: weeds. Proceedings of an international conference, Brighton, UK, 17-20 November 1997, British Crop Protection Council. pp. 1133-1140.
- Shaner, D. L.; Leonard, P. (2001). Regulatory aspects of resistance management for herbicides and other crop protection products. In Powles, S. B.; Shaner, D. L. Herbicide Resistance and World Grains. CRC Press, Boca Raton, FL. pp. 279-294. ISBN 9781420039085.
- Singh, D.V., 1993, Regional Specialization in the Cultivation of Off-Season Vegetables on Commercial Scale: A case of Himachal Pradesh, Agricultural Situation in India, Vol. 12, No. 13, 1993.
- Singh, D.V., 1997, Economics of Export Oriented Horticultural Crops of Himachal Pradesh, Agro-Economic Research Centre, Himachal Pradesh University, Shimla Sept.1997(Memio)
- Singh, Rajendra, 1997, Economics of Export Oriented Horticultural Crops in Uttar Pradesh, Agro-Economic Research Centre, University of Allahabad, Allahabad, 1997,(Memio).
- Stryer, Lubert (1995). Biochemistry, 4th Edition. W.H. Freeman and Company. p. 670. ISBN 978-0-7167-2009-6.
- Subrahmanyan, K.V., 1994, Impact of New Agricultural And Economic Policies on Export of Horticultural Produce From India, Indian Journal of Agricultural Marketing, Conf. Spl., 1994.
- Van Almsick, A. (2009). New HPPD-Inhibitors - A Proven Mode of Action as a New Hope to Solve Current Weed Problems. Outlooks on Pest Management. 20: 27-30. doi:10.1564/20feb09.
- Vats, S. (2015). Herbicides: history, classification and genetic manipulation of plants for herbicide resistance. In Lichtfouse, E. Sustainable Agriculture Reviews 15. Springer International Publishing. pp. 153-192.
- Weed Science Society of America.
- Zhou Q, Liu W, Zhang Y, Liu KK (Oct 2007). Action mechanisms of acetolactate synthase-inhibiting herbicides. Pesticide Biochemistry and Physiology. 89 (2): 89-96. doi:10.1016/j.pestbp.2007.04.004.

INDEX